Lithium-Ion Batteries

Safety Risks, Fire Hazards and Consumer Standards

Energy Science, Engineering and Technology Series

Human Capital, Energy and Sustainable Development: Global Empirical Modeling
Raufhon Salahodjaev
ISBN: 979-8-89530-444-0
eBook ISBN: 979-8-89530-512-6

Recent Trends in Energy Systems and Applications
R. Sivaraj, PhD,
Firuz Kamalov, PhD,
and Ho Hon Leung, PhD
ISBN: 979-8-89530-120-3
eBook ISBN: 979-8-89530-222-4

The Impact of Future Energy Sources on Geopolitical Reality
Medani P. Bhandari, PhD,
and Mohamad Mokdad, PhD
ISBN: 979-8-89113-936-7
eBook ISBN: 979-8-89530-017-6

Applied Geomechanics for the Analysis and Optimization of Cyclic Steam Stimulation Operations in Heavy Oil Reservoirs
Guillermo Arturo Alzate-Espinosa, MSc., Abel de Jesús Naranjo-Agudelo, BS., Edson Felipe Araujo-Guerrero, PhD, Carlos Andrés Torres-Hernández, MSc., Maria Camila Herrera-Schlesinger, BS., Camilo Andrés Benítez-Peláez, BS., Daniel Felipe Cartagena-Pérez, BS., Juan Camilo Correa-Álvarez, BS., and Elvis Fernando Higuita-Carvajal, BS.
ISBN: 979-8-89113-905-3
eBook ISBN: 979-8-89113-946-6

A Roadmap for Nuclear Waste Resolution
J. Jason Chao, PhD
ISBN: 979-8-89113-837-7
eBook ISBN: 979-8-89113-916-9

More information about this series can be found at
https://novapublishers.com/product-category/series/energy-science-engineering-and-technology/

Gregory R. Simmons
Editor

Lithium-Ion Batteries

Safety Risks, Fire Hazards and Consumer Standards

Library of Congress Cataloging-in-Publication Data

ISBN: 979-8-89530-773-1 (Softcover)
ISBN: 979-8-89530-949-0 (eBook)

Published by Nova Science Publishers, Inc. † New York

Contents

Preface

Fires in electric vehicles powered by high-voltage lithium-ion batteries pose a serious risk to emergency responders. These risks include electric shock from exposure to damaged high-voltage components and thermal runaway, where damaged cells experience uncontrolled increases in temperature and pressure. Thermal runaway can lead to hazards such as battery reignition or fires, driven by 'stranded' energy remaining in a damaged battery. Chapter 1 examines the safety risks emergency responders face when handling high-voltage lithium-ion batteries in electric vehicles. Chapter 2 reviews the fire hazards and safety concerns associated with lithium-ion batteries in everyday items like cell phones, computers, e-bikes, e-scooters, and electric vehicles. Chapter 3 proposes a consumer product safety standard for rechargeable lithium-ion batteries in micromobility devices.

Chapter 1

Safety Risks to Emergency Responders from Lithium-Ion Battery Fires in Electric Vehicles*

National Transportation Safety Board

National Transportation Safety Board. 2020. *Safety Risks to Emergency Responders from Lithium-Ion Battery Fires in Electric Vehicles.* Safety Report NTSB/SR-20/01. Washington, DC

Abstract

As part of an investigation into the safety risks emergency responders face when dealing with the high-voltage lithium-ion batteries that power electric vehicles, the National Transportation Safety Board (NTSB) studied three electric vehicle crashes resulting in fires (in Lake Forest and Mountain View, California, and Fort Lauderdale, Florida) and one noncrash fire involving an electric vehicle (in West Hollywood, California). The crashes caused extensive damage that extended into the protected area of the cars' high-voltage battery cases, rupturing the cases and damaging battery modules and individual cells. The noncrash fire was caused by an internal battery failure. In each case, emergency responders faced safety risks related to electric shock, thermal runaway, battery ignition and reignition, and stranded energy. The investigation

---

* This is an edited, reformatted and augmented version of Safety Report of National Transportation Safety Board, Publication No. NTSB/SR-20/01 PB2020-101011, Notation 59989, Adopted November 13, 2020.

In: Lithium-Ion Batteries
Editor: Gregory R. Simmons
ISBN: 979-8-89530-773-1

also examined national and international standards established to maximize the safety of electric vehicles and the emergency response guides produced by vehicle manufacturers. The NTSB identified two main safety issues: (1) inadequacy of vehicle manufacturers' emergency response guides for minimizing the risks to first and second responders (firefighters and tow operators) posed by high-voltage lithium-ion battery fires in electric vehicles, and (2) gaps in safety standards and research related to high- voltage lithium-ion batteries involved in high-speed, high-severity crashes. On the basis of its findings, the NTSB makes safety recommendations to the National Highway Traffic Safety Administration, the manufacturers of electric vehicles equipped with high-voltage lithium-ion batteries, and six professional organizations that represent or operate training programs for first and second responders.

The NTSB is an independent federal agency dedicated to promoting aviation, railroad, highway, marine, and pipeline safety. Established in 1967, the agency is mandated by Congress through the Independent Safety Board Act of 1974 to investigate transportation accidents, determine the probable causes of the accidents, issue safety recommendations, study transportation safety issues, and evaluate the safety effectiveness of government agencies involved in transportation. The NTSB makes public its actions and decisions through accident reports, safety studies, special investigation reports, safety recommendations, and statistical reviews.

The NTSB does not assign fault or blame for an accident or incident; rather, as specified by NTSB regulation, "accident/incident investigations are fact-finding proceedings with no formal issues and no adverse parties . . . and are not conducted for the purpose of determining the rights or liabilities of any person" (Title 49 *Code of Federal Regulations* section 831.4). Assignment of fault or legal liability is not relevant to the NTSB's statutory mission to improve transportation safety by investigating accidents and incidents and issuing safety recommendations. In addition, statutory language prohibits the admission into evidence or use of any part of an NTSB report related to an accident in a civil action for damages resulting from a matter mentioned in the report (Title 49 *United States Code* section 1154(b)).

For more detailed background information on this report, visit www.ntsb.gov and search for NTSB accident ID HWY19SP002. Recent publications are available in their entirety on the NTSB website. Other information about available publications may be obtained from the website or by contacting:

National Transportation Safety Board Records Management Division, CIO-40 490 L'Enfant Plaza, SW, Washington, DC 20594; (800) 877-6799 or (202) 314-6551

Copies of NTSB publications can be downloaded at no cost from the National Technical Information Service's National Technical Reports Library at https://ntrl.ntis.gov/NTRL/. This product may be accessed using product number PB2020-101011. For additional assistance, contact:

National Technical Information Service (www.ntis.gov/) 5301 Shawnee Rd. Alexandria, VA 22312; (800) 553-6847 or (703) 605-6000

Acronyms and Abbreviations

AC	alternating current
BEV	battery electric vehicle
CFR	*Code of Federal Regulations*
CTIF	International Association of Fire and Rescue Services (*Comité Technique International de prévention et d'extinction du Feu*)
DC	direct current
Euro NCAP	European New Car Assessment Programme
FMVSSs	*Federal Motor Vehicle Safety Standards*
GTR	global technical regulation
hazmat	hazardous materials
hp	horsepower
ISO	International Organization for Standardization
km/hr	kilometers per hour
kWh	kilowatt-hour
NCAP	New Car Assessment Program
NFPA	National Fire Protection Association
NFRD	Norwegian Fire and Rescue Department
NHTSA	National Highway Traffic Safety Administration
NiMH	nickel–metal hydride
NTSB	National Transportation Safety Board
PHEV	plug-in hybrid electric vehicle
PPE	personal protective equipment
SAE	SAE International (formerly Society of Automotive Engineers)
SCBA	self-contained breathing apparatus
SUV	sport utility vehicle
UNECE	United Nations Economic Commission for Europe
US-101	US Highway 101

Executive Summary

The National Transportation Safety Board (NTSB) investigated three electric vehicle crashes resulting in postcrash fires and one noncrash fire involving an electric vehicle, all of which illustrate the risks to emergency responders posed by the vehicles' high-voltage lithium-ion batteries. The NTSB also examined national and international standards established to maximize the safety of electric vehicles. Particular attention was given to the emergency guidance documents supplied by vehicle manufacturers to mitigate the safety risks to first and second responders who deal with electric vehicle crashes and high-voltage lithium-ion battery fires.[1]

Fires in electric vehicles powered by high-voltage lithium-ion batteries pose the risk of electric shock to emergency responders from exposure to the high-voltage components of a damaged lithium-ion battery. A further risk is that damaged cells in the battery can experience uncontrolled increases in temperature and pressure (thermal runaway), which can lead to hazards such as battery reignition/fire. The risks of electric shock and battery reignition/fire arise from the "stranded" energy that remains in a damaged battery.

Safety Issues

The investigation identified the following safety issues:

- Inadequacy of vehicle manufacturers' emergency response guides for minimizing the risks to first and second responders posed by high-voltage lithium-ion battery fires in electric vehicles.
- Gaps in safety standards and research related to high-voltage lithium-ion batteries involved in high-speed, high-severity crashes.

---

[1] Additional information related to this safety report (NTSB case number HWY19SP002) can be found by accessing the Docket Management System at www.ntsb.gov. For more information about NTSB safety recommendations, see the Safety Recommendation Database at www.ntsb.gov.

Findings

- Manufacturers' emergency response guides provide sufficient vehicle-specific information for disconnecting an electric vehicle's high-voltage system when the high-voltage disconnects are accessible and undamaged by crash forces.
- Crash damage and resulting fires may prevent first responders from accessing the high-voltage disconnects in electric vehicles.
- The instructions in most manufacturers' emergency response guides for fighting high-voltage lithium-ion battery fires lack necessary, vehicle-specific details on suppressing the fires.
- Thermal runaway and multiple battery reignitions after initial fire suppression are safety risks in high-voltage lithium-ion battery fires.
- The energy remaining in a damaged high-voltage lithium-ion battery, known as stranded energy, poses a risk of electric shock and creates the potential for thermal runaway that can result in battery reignition and fire.
- High-voltage lithium-ion batteries in electric vehicles, when damaged by crash forces or internal battery failure, present special challenges to first and second responders because of insufficient information from manufacturers on procedures for mitigating the risks of stranded energy.
- Storing an electric vehicle with a damaged high-voltage lithium-ion battery inside the recommended 50-foot-radius clear area may be infeasible at tow or storage yards.
- Electric vehicle manufacturers should use the International Organization for Standardization standard 17840 template to present emergency response information.
- Action by the National Highway Traffic Safety Administration, similar to that taken by the European New Car Assessment Programme, to incorporate scoring relative to the availability of a manufacturer's emergency response guide and its adherence to International Organization for Standardization standard 17840 and SAE International recommended practice J2990 into the US New Car Assessment Program, would be an incentive for manufacturers of vehicles sold in the United States with high-voltage lithium-ion battery systems to comply with those standards.

- Although existing standards address damage sustained by high-voltage lithium-ion battery systems in survivable crashes, as defined by federal crash standards, they do not address high-speed, high-severity crashes resulting in damage to high-voltage lithium-ion batteries and the associated stranded energy.

Recommendations to the National Highway Traffic Safety Administration

When determining a vehicle's US New Car Assessment Program score, factor in the availability of a manufacturer's emergency response guide and its adherence to International Organization for Standardization standard 17840 and SAE International recommended practice J2990. (H-20-30) Convene a coalition of stakeholders to continue research initiated by your organization on ways to mitigate or deenergize the stranded energy in high-voltage lithium-ion batteries and to reduce the hazards associated with thermal runaway resulting from high-speed, high-severity crashes. Publish the research results. (H-20-31)

To the manufacturers of electric vehicles equipped with high-voltage lithium-ion batteries: (BMW Group; BYD Motors; FCA Group; General Motors Company; Ford Motor Company; Gillig; Honda Motor Company; Hyundai Motor Company; Karma Automotive; Kia Motors Corporation; Mercedes-Benz USA; Mitsubishi Motors; Nissan Motor Company; Nova Bus, Inc.; Porsche Cars North America; Proterra Inc.; North American Subaru; Tesla, Inc., Toyota Motor North America; Van Hool NV; Volkswagen Group of America; and Volvo Car Corporation):

> Model your emergency response guides on International Organization for Standardization standard 17840, as included in SAE International recommended practice J2990, and incorporate vehicle-specific information on (1) fighting high-voltage lithium-ion battery fires; (2) mitigating thermal runaway and the risk of high-voltage lithium-ion battery reignition; (3) mitigating the risks associated with stranded energy in high-voltage lithium-ion batteries, both during the initial emergency response and before moving a damaged electric vehicle from the scene; and (4) safely storing an electric vehicle that has a damaged high-voltage lithium-ion battery. (H-20-32)

To the National Fire Protection Association, the International Association of Fire Chiefs, the International Association of Fire Fighters, the National Alternative Fuels Training Consortium, the National Volunteer Fire Council, and the Towing and Recovery Association of America:

> Inform your members about the circumstances of the fire risks described in this report and the guidance available to emergency personnel who respond to high-voltage lithium-ion battery fires in electric vehicles. (H-20-33)

1. Introduction

1.1. Overview

For more than 100 years, highway vehicles have been powered by internal combustion engines that burn gasoline or diesel fuel. In the last 20 years, however, vehicles running on *alternative fuels* have become popular, viewed as a means of reducing the world's dependence on fossil fuels, lowering fuel costs, and lessening environmental pollution. The most widely used alternative vehicle fuel is electricity.[2]

By 2000, vehicles combining an internal combustion engine with an electric motor, known as *hybrid-electric vehicles*, had been introduced to the US market (Honda Insight, 1999; Toyota Prius, 2000).[3] The vehicles used the electric motor as a secondary power source (to assist the internal combustion engine in acceleration, for example) and also to capture the kinetic energy from braking that would ordinarily be converted to heat and lost.[4] A nickel–metal hydride (NiMH) battery, charged by the internal combustion engine, powered the motors in the early hybrid vehicles.[5] The output of the motors was small, ranging from 13 to 40 horsepower (hp).

---

[2] Other common alternative vehicle fuels are natural gas, hydrogen, propane (liquefied petroleum gas), and biodiesel.

[3] The Prius was introduced in Japan in 1997; the Insight was launched in Japan the same year as in the United States (see Car and Driver website, accessed August 4, 2020).

[4] The process is known as *regenerative braking*. The captured energy is stored in the battery.

[5] Some manufacturers of electric vehicles, such as Toyota, continue to use NiMH batteries in their hybrid electric vehicles.

Plug-in hybrid electric vehicles (PHEVs) equipped with more powerful electric motors were introduced 10 years later.[6] The first widely available PHEV was the 2011 Chevrolet Volt. The Volt had a 149-hp electric motor powered by a high-voltage lithium-ion battery, in addition to a small, 84-hp gasoline engine.[7]

Today, *battery electric vehicles* (BEVs) are becoming the dominant alternative energy vehicles.[8] BEVs have a fully electric powertrain—that is, they do not have an internal combustion engine but are powered solely by an electric motor fueled by rechargeable batteries.[9] Early BEVs included the Tesla Roadster, introduced in 2008; the Nissan Leaf, introduced in 2010; and the Tesla model S, introduced in 2012. High-voltage lithium-ion batteries are the standard power source for BEVs.

1.2. Scope of Report

By virtue of its mandate to study transportation safety issues, the National Transportation Safety Board (NTSB) has an interest in the safety of emerging technology, including alternative fuel sources. Safety issues with the high-voltage lithium-ion batteries used in electric vehicles first gained widespread attention when a Chevrolet Volt caught fire 3 weeks after a crash test in May 2011.[10] The National Highway Traffic Safety Administration (NHTSA), which oversaw the crash test, investigated both the cause of the fire and the risk of fire in Volt cars that experienced serious crashes. After reviewing all reports of severe events involving Chevrolet Volts, NHTSA found no record of any crash-related fires involving batteries in a Volt, or in any other electric vehicle.[11] Although NHTSA found no defect with the Volt, Chevrolet

---

[6] PHEVs are so named because their batteries are charged by plugging into an external electrical socket.

[7] (a) In automotive engineering, *high voltage* is considered to be any amount above the maximum voltage safe for humans; see sections 1.3.1 and 1.3.2 for details. (b) The gasoline engine could generate power for the Volt's electric motor after the battery lost its charge, thereby extending the car's range.

[8] According to year-end data, BEVs accounted for three-quarters of electric vehicle sales in the United States during 2019 (combined sales of PHEVs and BEVs totaled 326,000). See Green Car Congress website and Argonne National Laboratory website, both accessed March 18, 2020.

[9] Batteries that produce the power to move a vehicle are called *traction batteries*.

[10] Section 3.1 describes the incident in detail.

[11] This information is from NHTSA's report on the Volt fire (Smith 2012). Severe events were considered those in which an airbag deployed, an occupant was injured, or the vehicle's

modified the design of the vehicle's battery case to offer improved resistance to crash damage.

At the time of the Volt fire, the transportation industry was beginning to evaluate the safety of lithium-ion batteries. In the weeks before the Chevrolet Volt fire, NHTSA sponsored a technical symposium on lithium-ion battery safety that summarized work by NHTSA, the US Department of Transportation, and the US Department of Energy in developing safer batteries for electric vehicles.[12] By then, NHTSA had developed a multitiered research plan to "understand failure risks, develop safety methods, and develop performance-based metrics" for lithium-ion batteries.[13] In the years that followed, NHTSA continued to sponsor conferences at which experts identified topics for further research.

In late 2011, NHTSA began working with the National Fire Protection Association (NFPA) to assist first responders (firefighters) and second responders (tow operators) in handling lithium-ion batteries after a crash, and was working with vehicle manufacturers to develop postcrash protocols for dealing with vehicles powered by lithium-ion batteries.[14] The previous year, the NFPA had published the results of a study of the hazards to firefighters and emergency responders from electric vehicles, though focus was on the NiMH batteries then in common use (Grant 2010). A more recent study (Long and others 2013) laid out best practices for emergency responders to hazards involving the batteries in electric vehicles. In 2015, the NFPA began publishing emergency field guides for alternative fuel vehicles as part of its safety training program.

The first NTSB investigations of lithium-ion battery fires involved safety risks in aviation. In 2013, the NTSB investigated a lithium-ion battery fire in a Boeing 787 aircraft in Boston andbassisted in investigating lithium-ion battery fires in another Boeing 787 aircraft (in Japan) and in a Boeing 747 (in the United Arab Emirates). As a result of its investigation of the Boeing 787 fire in Boston, the NTSB issued safety recommendations for assessing and

---

speed changed by more than 12 mph. See also NHTSA's press release of January 20, 2012, issued after its investigations were complete.

[12] The symposium (held May 18, 2011) was titled "Safety Considerations for EVs [electric vehicles] powered by Li-ion [lithium-ion] Batteries." The Department of Energy's slide presentation can be viewed on the NHTSA website (accessed November 12, 2020).

[13] As stated in Smith (2012), p. 4.

[14] (a) *First responders* in this context refers to firefighters, but emergency medical technicians, paramedics, and police officers are also classified as first responders. *Second responders* in this context refers to tow truck drivers or tow yard operators, but they can also include those responsible for temporary traffic control or other support functions at a crash site. (b) The NFPA's efforts are reported in Smith (2012).

managing the risk of short circuit and fire in lithium-ion batteries (NTSB 2014). Also in 2013, the NTSB convened a public forum titled "Lithium Ion Batteries in Transportation," and has continued to monitor safety issues related to lithium-ion batteries in aircraft. In 2018, the NTSB addressed hydrogen fuel cell electric vehicles when it investigated a fire on a vehicle transporting compressed hydrogen for use at a fueling station (NTSB 2019c). More recently, in May 2020, the NTSB issued a safety recommendation report titled Standards for Lithium-Ion Battery Shipments by Air (NTSB 2020b), following a fire involving large-format lithium-ion batteries that had been delivered by air. (See the appendix for further information on the above NTSB activities.)

The NTSB's first investigation of electric vehicle battery fires on US roadways was in 2017, when a high-voltage lithium-ion battery caught fire after a BEV left the road and crashed into a residential garage at high speed (for details, see section 2.1). By 2017, the market leader in the electric vehicle industry was Tesla, Inc., which now accounts for about 80 percent of BEV sales in the United States.[15] Between 2017 and 2018, the NTSB investigated two other high-speed, high-severity crashes that resulted in postcrash fires (see sections 2.2 and 2.3) and one noncrash fire involving a BEV (see section 2.4), all of which involved Tesla-manufactured vehicles. During the course of its investigations, the NTSB considered the safety risks to first and second responders posed by the vehicles' high-voltage lithium-ion batteries—risks that differ from the issues presented by fires in vehicles powered by internal combustion engines. This safety report focuses on those risks.

As the NTSB concluded its investigations, international incidents concerning other vehicle manufacturers came to light, including the three high-voltage lithium-ion battery fires in Europe that are described in section 3. In addition, NTSB investigators evaluated six data sources for information on the prevalence of fires in electric vehicles. The research found that while each database contained some information related to vehicle fires, all had serious limitations—for example, in their ability to identify batteries as the source of vehicle fires.[16]

Section 4 describes regulatory and industry actions undertaken to maximize the safety of both the vehicles powered by high-voltage batteries

---

[15] As reported for 2017 by Statista, "Tesla Dominates the US Electric Vehicle Market," and for 2019, also by Statista, "Estimated Battery Electric Vehicle Sales in the US by Brand 2019" (both accessed September 28, 2020).

[16] The data report ("Prevalence of Electric Vehicle Battery Fires") is available in the NTSB public docket (case number HWY19SP002).

and the emergency personnel who respond to crashes and other incidents involving those vehicles. The section concludes with a summary of the guidance available to emergency responders, including a detailed review of the emergency response guides published (voluntarily) by the manufacturers of electric vehicles.[17] Section 5 presents the NTSB's analysis and makes recommendations for addressing (1) inadequacies in the vehicle manufacturers' guidance for minimizing the risks to emergency responders posed by high-voltage lithium-ion battery fires, and (2) gaps in the standards and research related to high-voltage lithium-ion batteries involved in high-speed, high-severity crashes. Below, we characterize the lithium-ion batteries that power BEVs and outline the unique risks they pose to first and second responders.

1.3. High-Voltage Lithium-Ion Batteries

BEVs require high-energy batteries—batteries that store significant quantities of energy (electricity), retain it efficiently, and discharge it at a high rate. Lithium-ion batteries have been chosen for BEVs because they have high energy density (allowing them to store large amounts of energy for a given volume), a low self-discharge rate (allowing them to retain a charge), and excellent electrochemical potential (allowing high-power discharge). Protection circuits are necessary to maintain charging and discharging within safe limits.

1.3.1. Properties

Like all batteries, lithium-ion batteries consist of cells that produce an electric current by converting chemical energy into electrical energy. Each battery cell consists of two electrodes (one negative, the anode, and one positive, the cathode) that conduct electricity. A permeable barrier between the electrodes prevents internal short-circuiting and allows another substance, the electrolyte, to transfer charged ions between the electrodes.[18] The anode is typically made of carbon (graphite), and the cathode generally consists of layers of lithium

[17] There are no federal requirements for emergency response guides, although regulations cover the transportation and disposal of high-voltage lithium-ion batteries.

[18] The *charged ions* in a lithium-ion battery are atoms of the element lithium that are missing one electron, giving them a positive electrical charge. Lithium is the lightest metal on the periodic table of elements.

and metal oxide.[19] The electrolyte consists of a lithium salt dissolved in an organic solvent, mainly carbonates. (Electrolytes in lithium-ion batteries can be liquids or gels.) The organic solvents used in lithium-ion batteries are flammable.[20]

The more ions an electrode can absorb and release in relation to its size and weight—known as capacity—the more energy it can store. Capacity is measured in kilowatt-hours (kWh)— meaning the amount of electricity a battery can deliver or absorb over the course of 1 hour. The battery in the 2011 Chevrolet Volt (a PHEV equipped with both an internal combustion engine and an electric motor) has a capacity of 16 kWh. In the fully electric 2019 Tesla model S, which has both a basic and a high-performance version, the battery capacity ranges from 60 to 100 kWh.

The voltage of a battery (expressed in volts) is its potential to produce an electric current— that is, to push electrons around a closed circuit.[21] The lithium-ion batteries in electric vehicles are described as high-voltage because they have a potential of 300 to 400 volts or more, with 800-volt electric vehicles planned or already on the market.[22] In comparison, the familiar lead-acid car battery, which supplies power to start an internal combustion engine and to run a vehicle's auxiliary electrical systems, has just 12 volts.

Lithium-ion battery cells come in various shapes and sizes (cylindrical, prismatic, elliptical, pouch). The battery for a BEV consists of individual cells packed tightly together to produce the required voltage, power, and energy. The cells are assembled into modules, and the modules assembled into battery packs and systems. The battery is packaged in a case designed to resist damage from external forces and located for protection from crashes.

Battery management systems maintain the safe operation of battery packs. In addition to monitoring voltage and temperature data from the cells and modules, they monitor the battery's state-of-charge (level of charge relative to capacity) to protect against overcharging or undercharging. The protection

---

[19] Various metal oxides are employed in lithium-ion batteries, including lithium cobalt oxide (LiCoO2), the most common. The batteries in the four fires investigated by the NTSB contained lithium nickel manganese cobalt oxide (LiNiMnCoO2). See the online Battery University ("BU-205: Types of Lithium-ion") for more information (accessed November 12, 2020).

[20] Researchers have concluded that the flammable solvents in lithium-ion batteries are about as hazardous as the gasoline or diesel used in conventional vehicles (Stephens and others 2017, pp. 2-30–2-31).

[21] Technically, voltage is the difference in electric potential between two places, called the *electric potential difference.*

[22] Porsche launched an 800-volt BEV sports car, called the Taycan, in 2019, as reported on the Green Car Congress website (accessed April 24, 2020).

circuits mentioned earlier limit peak cell voltage and prevent the voltage from dropping too low. The cells are protected from temperature extremes by thermal management systems integrated into the modules and from overpressure by venting systems. Some battery designs incorporate liquid cooling systems.

Different vehicle manufacturers use different batteries and different battery packs. As shown in figure 1, the Tesla model S (like other of the manufacturer's vehicles) has a flat battery pack that lies under the floor. The battery pack is made up of thousands of small, cylindrical battery cells (slightly larger than an AA battery) packaged into 16 modules. The Chevrolet Volt, as an example from another manufacturer, uses pouch cells (about the size of a standard 8 1/2-inch by 11-inch envelope), and the battery pack is T-shaped, with the top of the T fitting under the car's back seat. The Volt's battery pack contains 192 cells packaged into four modules.

Lithium-ion batteries produce direct current (DC) power. Owners of electric vehicles can charge their cars by plugging them into a 120- or 240-volt alternating current (AC) residential outlet because the cars are equipped to convert AC power to DC. The batteries can also be charged at commercial DC fast-charging stations.[23] The stations can charge a vehicle in less than an hour, compared with several hours using a residential outlet.

1.3.2. Safety Risks

Fires in electric vehicles powered by lithium-ion batteries pose two main dangers to emergency responders. First is the risk of electric shock from exposure to high-voltage connections in a damaged battery. Second is the risk that damaged cells in the battery will experience uncontrolled increases in temperature and pressure, known as thermal runaway, which can lead to venting and combustion of toxic gases, cell rupture and release of projectiles, and battery reignition/fire. The risks of electric shock and battery reignition/fire arise from the energy that remains in a damaged battery—known as stranded energy.

---

[23] The EVgo network installs DC fast-charging stations at such locations as grocery stores (accessed November 12, 2020).

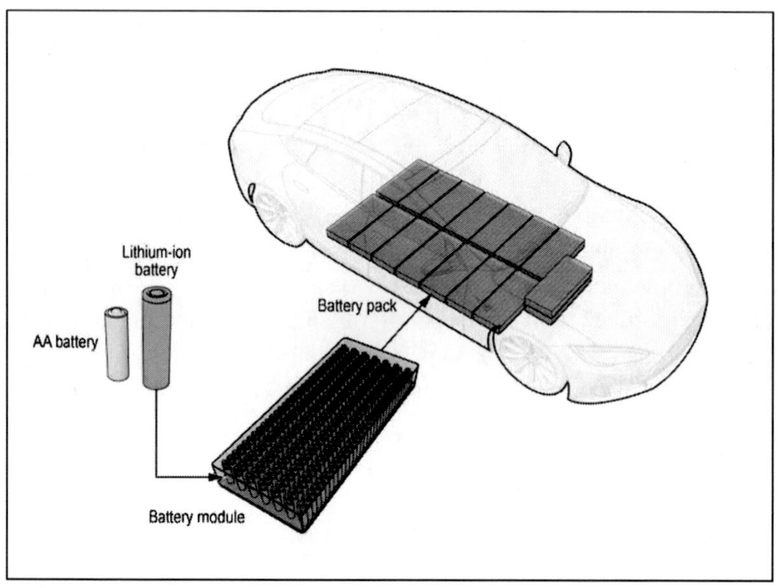

Figure 1. Illustration of Tesla model S showing location of battery pack and details of module and battery cell, with size comparison to standard AA battery. (NTSB drawing).

Electric shock. The human body is an electrical conductor; if it contacts an energized source of electricity, current will flow through it. The body's *resistance*—its ability to reduce an electric current—varies from person to person and according to whether the skin is wet or dry, among other things. The maximum voltages considered safe for humans are 50 or 60 volts DC and 30 volts AC.[24] The high-voltage system of a BEV operates well above those thresholds (at 300 to 400 volts or more), creating a safety risk when the high-voltage battery is damaged in a crash and safety features such as protective covers and circuit fuses are defeated. As described in section 4, to protect occupants, bystanders interacting with injured persons, and emergency responders from electric shock, safety standards require electrically isolating the high-voltage battery system from the vehicle's chassis. If a crash damages the electrical isolation system, a person who touches the vehicle (or an exposed

[24] The thresholds are based on International Electrotechnical Commission technical standard 60479-1; 2005-07, "Effects of Current on Human Beings and Livestock." The physiological effects of an electric current can be anything from a slight prickling sensation to burns and respiratory or cardiac arrest. DC current has a less-severe effect on the human body than AC current.

connector) can become part of the high-voltage circuit and suffer serious injury or death.

Thermal runaway. Thermal runaway is a chemical process that produces heat (an *exothermic reaction*); the heat increases the rate of the reaction, which further increases the temperature and escalates the process. Thermal runaway can spread from one battery cell to many cells, in a domino effect. The originating cause of thermal runaway is generally short-circuiting inside a battery cell and a resulting increase in the cell's internal temperature. A short circuit in a lithium-ion battery cell can result from defects introduced during manufacturing, such as contamination, or from damage to the cell caused by crushing or puncturing—precisely the kind of damage produced by high-impact, high-severity car crashes. An external fire might also heat a battery cell enough to initiate thermal runaway.

Fire and explosion can result when cells go into thermal runaway. The flammable solvent in the electrolyte can ignite if exposed to high temperatures or electrostatic sparks. Popping or other noises resulting from the venting of heat and gases often accompany a thermal runaway. A recent study (Stephens and others 2017) identified four primary hazards of thermal runaway:

(1) venting of toxic and flammable vapors from the electrolytic solvent, through pressure-relief devices or holes in the battery casing; (2) combustion of vapors ejected from the flammable electrolyte solvent; (3) localized overpressure; and (4) rupture of the cell casing and release of projectiles if pressure-relief devices are absent or fail. Secondary hazards identified in the study were release of toxic and corrosive chemicals, ignition and burning of combustible parts of the vehicle, asphyxiation of vehicle occupants from toxic gases vented by the battery, and electric shock to occupants, first responders, or maintenance personnel from exposure to high-voltage conductors if electrical insulation and isolators melt or burn. Flammable gases (hydrogen, ethylene, ethane, and propane) released from a damaged battery constitute the most significant fire threat, according to the study.

As detailed in later sections of this report, extinguishing a burning lithium-ion battery can require applying thousands of gallons of water. Section 3 describes a method that firefighters in Europe have used to extinguish a lithium-ion battery fire—submerging the entire vehicle in a vat filled with water. That method, however, creates a potential for short-circuiting and might also be impractical.

Stranded energy. If a high-voltage battery is damaged, energy remains inside any undamaged battery modules and cells, with no path to discharge it. That stranded energy can cause a high-voltage battery to reignite multiple times after firefighters extinguish an electric vehicle fire. Emergency responders have no way of measuring how much energy remains in a damaged battery, and no way of draining that energy, other than such time-consuming methods as allowing a battery fire to burn itself out.[25] Engineers or other specialists can use the battery management system to check for remaining voltage if the system is operational, and some batteries have built-in discharge ports, also for use by specialists. However, the high-voltage battery system can be damaged in a crash, preventing access to the battery management system or to the discharge ports. Moreover, as described in section 4, one of the first steps in responding to an electric-vehicle fire is to cut the supply cable to the 12-volt battery—which will depower the battery management system.

Electric vehicles are often equipped with emergency cut loops, low-voltage wire loops that first responders can safely cut to disconnect the high-voltage system from the rest of the vehicle.[26] Severing the cut loops will isolate high-voltage power inside the battery, thereby protecting the rest of the vehicle. However, cutting the loops will not remove energy from the high-voltage battery.

Manufacturers have developed tools to drain the high-voltage batteries in their vehicles, but the tools, which require a specialist to operate, are usually specific to the vehicle and work only on an intact battery.[27] One method of deenergizing a damaged battery is to submerge it in a saltwater bath (saltwater conducts electricity). It might not be possible, however, to extract a damaged battery from a vehicle after a severe crash, and first responders generally lack the expertise to remove a damaged battery.

2. NTSB Investigations of Electric Vehicle Battery Fires

The NTSB investigated all US domestic high-voltage battery fires in electric vehicles that we became aware of during a one-year period (August 2017 to

---

[25] The modules in Tesla batteries are separated in such a manner that allowing one of those batteries to burn itself out would be unlikely to remove all the energy.

[26] The wires are tagged at the location where they should be cut.

[27] Tools and techniques for assessing stranded energy are investigated in Rask and others (2020). The authors developed a prototype discharge tool, suitable for use by nonexperts, that would connect to a dedicated high-voltage access port in a protected part of an electric vehicle. The tool would, however, require access to a functional battery management system or direct connection to internal battery modules. Such access would be problematic, or impossible, in the case of a battery that had been damaged in a crash or had suffered thermal runaway.

August 2018).[28] In the four fires involving lithium-ion batteries that the NTSB identified during that period, three of the batteries had been damaged in high-speed, high-severity crashes that preceded the fires. All three crash-damaged batteries reignited after firefighters extinguished the vehicle fires. (For purposes of this report, a battery fire reignition is defined as a fire event—smoke, popping noises from the battery, or actual flames—that occurs in the battery after the vehicle fire has been extinguished and the vehicle has been stable for several minutes.)[29] The battery in the fourth case—a fire that occurred during normal vehicle operations—did not reignite.

All the batteries were examined after the incidents, as described below. For each incident, we indicate whether first and second responders consulted the emergency response guidance made available by the vehicle maker or contacted the manufacturer directly.[30] Note that full personal protective equipment (PPE) and self-contained breathing apparatus (SCBA) are standard firefighter equipment for all vehicle fires, not just those involving electric vehicles (Long and others 2013, p. 18).

2.1. Lake Forest, California, August 2017

On Friday, August 25, 2017, at 6:17 p.m. Pacific daylight time, a 2016 Tesla model X sport utility vehicle (SUV), occupied by a driver and one passenger, was traveling on a residential street in Orange County, California, that had a posted speed limit of 35 mph. The driver lost control of the vehicle, which left the road, crossed a sidewalk and an embankment, traveled down a drainage ditch, and collided with a culvert. Ultimately, the SUV hit a property wall, an open garage, and an unoccupied car parked in the garage. According to the SUV's onboard event data recorder, the driver had accelerated to 82 mph shortly before the crash.[31] The driver sustained serious injuries in the crash, and the passenger received minor injuries. A postcrash fire spread from the

[28] All the vehicles were BEVs.

[29] Reignition can occur with lithium-ion batteries because the battery itself contains the three elements necessary to sustain a fire (heat, fuel, and oxygen). Popping noises inside or between battery cells indicate an exchange of energy.

[30] See sections 4.3 and 4.4 for recommended practices and guidance for responding to electric vehicle fires. Emergency response guides are discussed in section 4.5.

[31] A report on the data retrieved from the vehicle's electronic control unit can be found in the public docket for this crash (NTSB case number HWY17FH013). The Factual Report of Investigation and other documents in the public docket give further details about the crash.

SUV to the car parked in the garage, the garage itself, and the house (figure 2).

Source: Orange County Sheriff's Department).

Figure 2. Postcrash view of garage showing firefighter directing water onto burning SUV and smoke coming from garage roof.

2.1.1. Initial Response

The Orange County Fire Authority received the first alarm at 6:17 p.m., and firefighters arrived on scene at 6:25 p.m. Officers from the Orange County Sheriff's Department arrived at 6:28 p.m. and set up a unified command. The officers managed the scene, which included identifying those involved and taking witness statements.

When firefighters arrived, both the SUV and the house were on fire. The first crew on scene attacked the vehicle fire in the garage with water. By 6:44 p.m., the bulk of the fire was out, but there appeared to be a fuel source in the garage, and fire was burning in the attic above the garage. The garage structure was sagging, making it difficult for firefighters to access the attic fire, which threatened the house. Firefighters halted the progress of the structural fire by about 7:00 p.m. The fire under the SUV appeared to be extinguished but then reignited an unspecified number of times.

By 7:17 p.m., the crew from the heavy rescue truck on scene had stabilized the garage wall. Firefighters then removed the SUV from the garage to assess the fire and identified the fuel source as the SUV's high-voltage battery pack. At 8:04 p.m., after the SUV had been pulled onto the driveway, the battery reignited. The fire was quickly extinguished using water, after which the SUV remained stable for about 45 minutes. During that time, firefighters brought the house fire under control.

The SUV began to emit heavy white smoke 45 minutes after the flames had been extinguished. The SUV ignited again and began burning in what firefighters described as a "blowtorch" manner. Firefighters applied water at up to 200 gallons per minute, but that did not extinguish the flames. The crew allowed the vehicle to burn freely to eliminate as many interior combustibles as possible. At 9:13 p.m., firefighters propped the SUV on cribbing blocks to expose the underside and applied more water, at a maximum rate of 600 gallons per minute, for about 45 minutes to cool the battery.[32] Applying water to the underside of the SUV extinguished the fire.

One of the responding battalion chiefs told investigators that the fire was "very severe" and more difficult to extinguish than firefighters expected. Even though they applied a large amount of water, the underside of the vehicle kept reigniting and would not go out. According to the battalion chief, the fire crews reviewed emergency response guides and searched online for guidance on the appropriate action to take.[33] The chief said that firefighters used breathing apparatus because of the large amounts of acrid smoke—he said it felt almost like a hazmat (hazardous materials) fire. He also said that those commanding the fire response would have liked to let the vehicle fire burn itself out, but they were concerned that it could take up to 24 hours and that the smoke would affect people in the neighborhood.[34]

As a result of the initial emergency response efforts, two Orange County Sheriff's Department officers sustained minor injuries from smoke inhalation. More than 20,000 gallons of water were applied to the vehicle fire, at varying rates, over at least 2 hours.

---

[32] *Cribbing blocks* are temporary wooden structures used to support heavy objects during construction, vehicle extrication, and so forth.

[33] The chief did not say explicitly that the crew consulted the vehicle manufacturer's emergency response guide, but the guide for the SUV gives locations and descriptions of high-voltage components, airbags, inflation cylinders, seat belt pretensioners, and the high-strength materials in the vehicle body. The guide includes a high-voltage disabling procedure and safety considerations specific to this SUV. See section 4.5 for more information about the emergency response guides and how they can be accessed.

[34] The vehicle's emergency response guide states: "Battery fires can take up to 24 hours to extinguish. Consider allowing the battery to burn while protecting exposures."

2.1.2. Secondary Response

A vehicle towing service was called at 10:40 p.m. and arrived at 11:09 p.m. While the SUV was being loaded onto the tow truck, it again began to emit smoke, and the battery fire reignited (figure 3). Firefighters applied water, but they had difficulty directing the water under the vehicle because it was resting on planks (its wheels had been displaced in the crash). The SUV was lowered off the flatbed, and firefighters again extinguished the fire. The tow truck driver suffered minor burns to his arms while lowering the SUV off the flatbed because the controls were near the sides of the burning vehicle, and he had to remove his wet, slippery gloves to operate them.

Firefighters continued applying water at about 300 gallons per minute to cool the battery. Once the battery had cooled, the SUV was towed from the scene. The time was 12:21 a.m., about 6 hours after the crash.

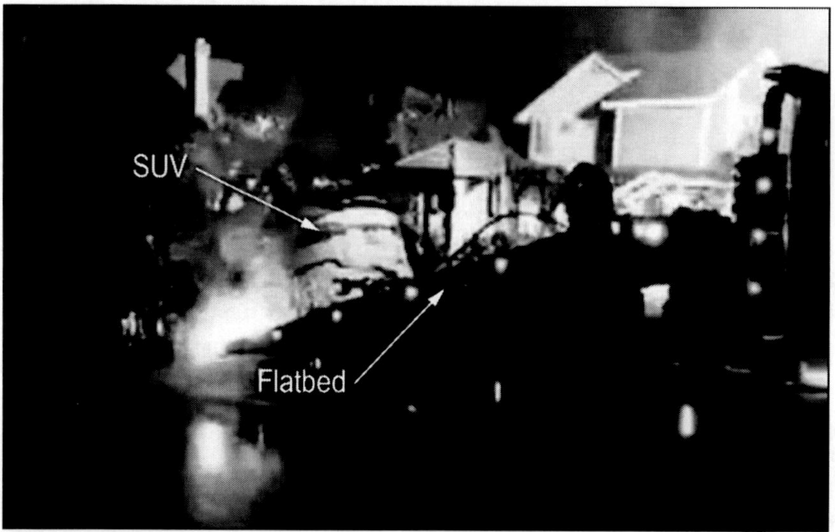

Source: Orange County Sheriff's Department.

Figure 3. Still image from video taken by sheriff's department officer showing SUV reignited on flatbed of tow truck.

The battery reignited for a few seconds while the SUV was being unloaded at the tow yard. Although the vehicle emitted smoke, it did not catch fire, and the tow driver did not call for an emergency response.[35] The SUV

[35] As noted earlier, a fire event can consist of smoke or popping noises, not necessarily involving flames.

was positioned at the tow yard as far from other vehicles or buildings as possible—about 40 feet from other vehicles on two sides, about 20 feet from other vehicles on the third side, and about 10 feet from a concrete wall on the fourth side. According to the towing log, the job was completed at 1:15 a.m.

2.1.3. Postcrash Inspection

On September 6–7, 2017, NTSB investigators examined the SUV and the battery at the tow yard. They also visited the crash site. The SUV had extensive fire and impact damage. The high-voltage cut loop had been in the area of severe fire and impact damage to the interior, making it impossible to determine whether firefighters had been able to access and cut it. When the vehicle was raised so the underside could be viewed, investigators found that the right front corner of the battery case had ruptured, revealing individual battery cells, as shown in figure 4.

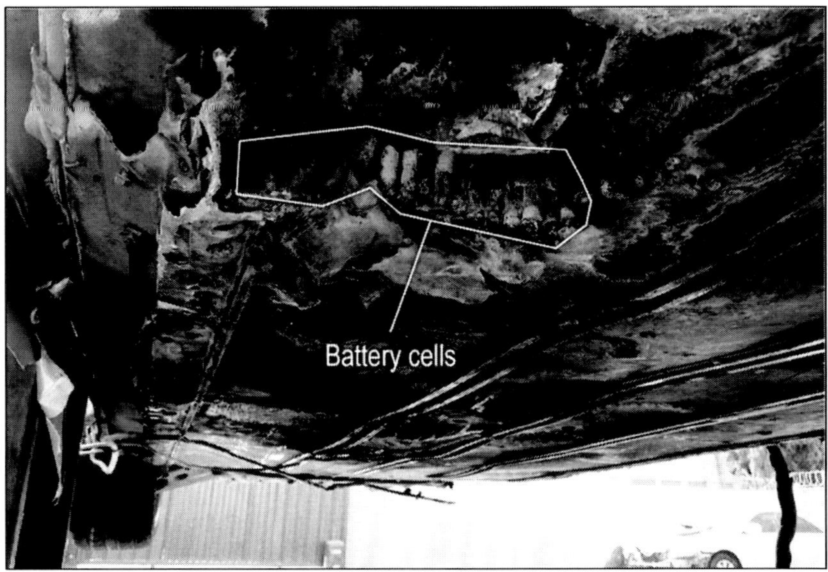

Figure 4. View from below and toward rear of SUV's underside showing ruptured right front corner of battery case and highlighting individual battery cells uncovered by rupture. Right rear wheel is visible in background.

The SUV's lithium-ion battery contained 16 modules (figure 5).[36] The rupture exposed parts of modules 13 and 15 at the right front corner, in line with where the right front wheel, brake disk, and other components had been displaced, most likely in the SUV's collision with the drainage culvert. Modules 11 and 9, directly behind module 13, were damaged. Battery cells from modules 13 and 15 had been dislodged or displaced, and individual cells were found on the street, in the driveway, and in the drainage ditch.[37]

The battery showed thermal damage, warping, and deformation to about 80 percent of the exterior of the battery case (refer to figure 4). Only the rear corners did not exhibit thermal damage. All the overpressure valves showed signs of hot gas venting, and the vents at the front were completely burned away.[38] Those on the left side and in the middle showed extensive thermal damage, while those in the rear had less.

Although the battery exhibited thermal damage, much of it remained intact. The duration of the postcrash fire and the multiple battery reignitions were evidence that the battery contained stranded energy. The amount of stranded energy could not be directly measured, however, because the module connection terminals were inaccessible.

2.2. Mountain View, California, March 2018

On March 23, 2018, at 9:27 a.m. Pacific daylight time, a 2017 Tesla model X SUV was traveling on US Highway 101 (US-101) in Santa Clara County, California. The SUV entered a paved gore area dividing the main travel lanes of US-101 from State Route 85 and struck a damaged, nonoperational crash attenuator at the end of a concrete barrier at 71 mph.[39] The SUV collided with

---

[36] All the batteries discussed in this section of the report had the same basic configuration. Readers are invited to refer to figure 5 in the sections where battery damage in the other vehicles is described.

[37] Each module in the SUV's battery had 515 cells.

[38] The spaces between the modules of the battery case contained rows of small vents housing one-way valves that would discharge hot gas in the event of thermal runaway. The valves were positioned to direct the venting gas away from the passenger compartment.

[39] A *crash attenuator* is a type of traffic safety hardware designed to protect motorists by reducing the collision forces on a vehicle. When the front of a vehicle hits an attenuator, the device telescopes to the rear and helps absorb the colliding vehicle's impact energy. Attenuators are usually placed in front of fixed structures on highways, such as barriers that separate traffic lanes. See the NTSB's safety recommendation report related to the collision with the attenuator in Mountain View (NTSB 2019a).

two other cars and caught fire after coming to rest (figure 6). The driver died of traumatic injuries sustained in the crash.[40]

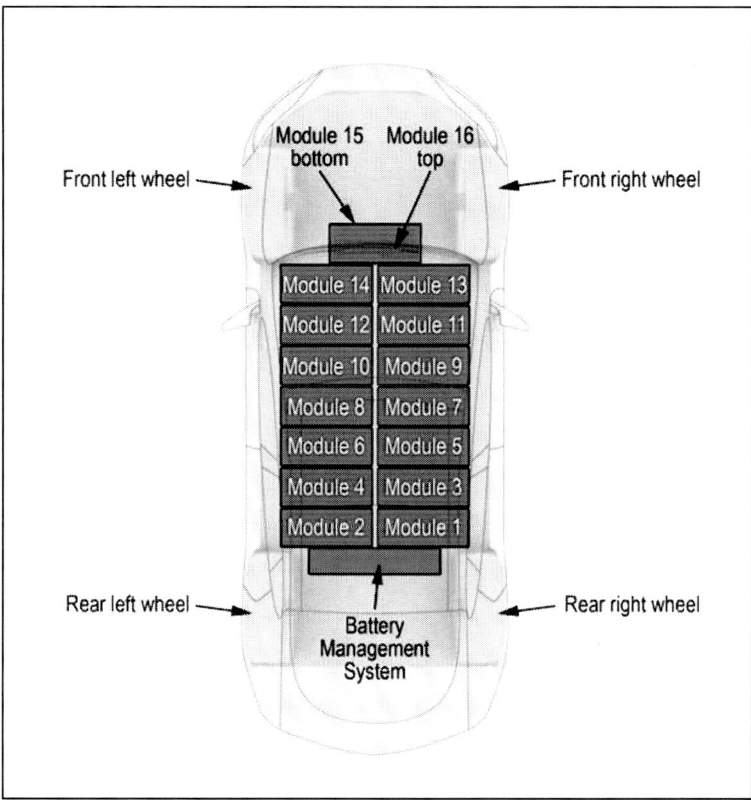

Figure 5. Arrangement of SUV's lithium-ion battery modules.

---

[40] For further details, see the NTSB's report on its investigation of the Mountain View crash (NTSB 2020a).

Source: witness S. Engleman.

Figure 6. Burning SUV on northbound US-101 postcrash.

2.2.1. Initial Response

The first emergency 911 call was received at 9:28 a.m. Three minutes later, the Mountain View Fire Department dispatched equipment and personnel to the crash scene (three engine units, one rescue unit, and a truck for patient care, fire suppression, and equipment or personnel recovery). The fire apparatus and rescue units arrived between 9:37 and 9:40 a.m. The California Highway Patrol joined the on-scene response at 9:46 a.m. and coordinated the work.[41]

Firefighters told NTSB investigators that the vehicle fire was extinguished quickly, using a mixture of water and foam. They estimated that suppressing the vehicle fire took only about 30 seconds for the front and a similar time for the rear, using about 200 gallons of water and foam. Electrical arcing was visible at the front of the vehicle during fire suppression. After the flames had been extinguished, intermittent popping noises were heard, accompanied by smoke, which prompted firefighters to apply more water. Exposed electrical

[41] Further documentation of the emergency response can be found in the Vehicle and Survival Factors factual report in the public docket for the Mountain View investigation (NTSB accident ID HWY18FH011).

cables were visible in the wreckage. Firefighters tried to use a "hot stick" to measure voltage on the vehicle surfaces and exposed cables, but it did not yield a reading because it was designed to detect AC voltage.[42]

Because of concerns about high voltage and stranded energy associated with the lithium-ion battery, the fire department's incident commander contacted the manufacturer about additional actions needed to make the vehicle safe. One of the manufacturer's battery engineers advised that the vehicle was not safe because of the extent of damage and that all personnel should stay away from the SUV until it could be evaluated by the manufacturer's representatives. (The crash occurred not far from the manufacturer's headquarters and its factory.)

The fire department's battalion chief coordinated with the California Highway Patrol, which agreed to keep the highway closed and wait for support from the manufacturer. While emergency responders waited, a loud popping sound came from the SUV, but no fire was apparent, so no further fire suppression was undertaken. The manufacturer sent two battery engineers, who arrived on scene about 12:30 p.m. They attempted to remove the damaged battery parts and placed loose battery cells and other battery components, including module 16 (which was in the wreckage but dislodged from the battery case) in a large bucket half-filled with water.

While the engineers attempted to remove the damaged battery components, popping sounds again came from the vehicle, and the floor of the SUV was observed to shift. The engineers moved away from the vehicle and determined that further attempts to remove damaged parts would be unsafe. A large part of the battery appeared to be intact, which made the battery a high-voltage safety risk.

The engineers determined that the SUV could not be stabilized on scene and that it should be removed. The engineers said that it would be best to maintain the vehicle on a flat surface to avoid flexing the vehicle structure (which could reignite the battery), but because the bed of the tow truck that came to the scene was metal, wooden blocks were used to maintain electrical isolation. The engineers advised having a fire truck escort the tow truck, despite anticipating an hour-long journey to the tow yard, so an escort was provided. The SUV was loaded onto the tow truck, and the scene was cleared at 3:05 p.m.

---

[42] A hot stick is safety device consisting of an insulated pole that responders can use to contact energized objects from a safe distance. New technology is being developed to measure DC voltage (Kane 2018).

2.2.2. Secondary Response

The tow truck carrying the SUV was escorted by the California Highway Patrol and the Mountain View Fire Department to an impound yard in San Mateo, California. The engineers followed. The SUV arrived at the yard at 4:17 p.m. The engineers told NTSB investigators that they stressed the importance of leaving a 50-foot radius around the vehicle, as recommended by the manufacturer's emergency response guide, but that there was not enough room in the yard, so workers did the best they could to separate the SUV from objects that appeared to be flammable.

According to the engineers, about 20 minutes after the SUV arrived at the impound yard, a California Highway Patrol officer heard popping sounds coming from the wreckage and called the San Mateo Fire Department. A fire engine arrived at 4:46 p.m. Firefighters monitored the SUV with thermal cameras, but they took no fire suppression action. The fire engine was dispatched again to the impound yard less than an hour later (at 5:20 p.m.), after the battery reignited. Crews monitored the vehicle, but the fire went out on its own and firefighters took no fire suppression action.[43]

The battery reignited once again 5 days after the crash. At 7:01 p.m. on March 28, a security guard reported a fire at the impound yard. He had seen smoke coming from under a tarp that had been placed over the SUV wreckage. California Highway Patrol investigators had inspected the vehicle earlier that day, and in the course of removing electronic equipment, had stood on parts of the vehicle near the front of the battery case.

According to the San Mateo Fire Department's dispatch log, firefighters arrived at 7:09 p.m. and reported flames 8 to 12 inches high coming from the right front side of the SUV. Firefighters suppressed the fire using water and foam. A security camera recorded a firefighter applying water and foam onto the SUV, which was still partly covered by the tarp and a road sign that had been used to anchor the tarp (figure 7).

A hose with a capacity of 90 gallons per minute was used until it depleted the engine's 500 gallons of water, after which firefighters laid a supply line to a fire hydrant. They applied an estimated 600 to 700 gallons of water over 30 to 40 minutes, but the fire continued to emit smoke and burn. The battalion chief on scene told NTSB investigators that firefighters were concerned about the risk of electricity traveling up the water stream and had been cautiously

---

[43] The small lithium-ion battery cells used in the battery were designed to create robust separation between cells. In this case, that design was a factor in the reignitions having stopped without further intervention and without consuming the rest of the battery.

starting and stopping the application of water.[44] Firefighters telephoned the manufacturer's engineers, who advised them to apply foam and that electricity "should not be a big risk," in the chief's words. At 8:10 p.m., after about 5 minutes of foam application, no further smoke or fire was observed.

Source: Atlas Towing.

Figure 7. Image from security video at impound yard showing firefighter pouring water onto SUV at right, partly covered by tarp and road sign; fire engine is visible on left.

Firefighters monitored the vehicle with a thermal camera (they also checked the temperature in the bucket of water containing battery parts assembled on scene). An engineer from the manufacturer arrived and monitored the vehicle for additional smoke, but none appeared. A San Mateo Fire Department hazmat unit tested the fire runoff and determined that it was toxic. A public works crew was brought in to vacuum about 600 gallons of water, foam, and vehicle fluids from the nearest storm drain.[45] The scene was declared safe at 9:50 p.m.

Combining both initial and secondary responses, a total of about 1,400 gallons of water and foam was used to extinguish the battery fires.

---

[44] In 2013, the NFPA commissioned research (reported in Long and others 2013) on the potential for firefighters to suffer electric shock from the water stream used to suppress electric vehicle fires—a concern because water is an electrical conductor. The research showed that the electrical current between the vehicle chassis and the firehose nozzle used in the tests was negligible, as were voltage levels at the nozzle. No adverse electrical conditions were noted. As a result of the tests, the NFPA's emergency field guides (see NFPA 2018) include a statement that "The use of water or other standard agents does not present an electrical hazard to firefighting personnel."

[45] A San Mateo environmental health and hazardous materials specialist consulted with the fire department in determining what actions to take.

2.2.3. Postcrash Inspection

Investigators from the NTSB and representatives of the California Highway Patrol inspected the SUV on March 27 and 28, 2018, and again on April 6, 2018. They were joined at the impound yard on April 6 by a fire engine and crew from the San Mateo Fire Department and by engineers from the manufacturer, who attempted to deenergize the vehicle's high-voltage lithium-ion battery (remove its stranded energy).

According to discussions with the manufacturer, the SUV's high-voltage battery had a discharge port under the battery case, covered by a panel, which could be used to discharge the battery. To access the port, the engineers had the vehicle elevated on wooden blocks, but only high enough to access the port so as to minimize vehicle movement. The underside of the SUV appeared undamaged by physical or thermal effects.

Engineers removed the panel, then pulled the plug from the discharge port. Water and debris drained from the port, and mud and debris could be seen on the inside of the cover. The water and debris had rendered the port's electrical connector nonfunctional. Engineers tried to connect a discharge apparatus to the exposed battery terminals at module 14, at the front of the ruptured battery case (refer to the battery diagram in figure 5, section 2.1.3). The discharge apparatus's computer showed that the terminal was not connected to a functioning battery, which meant that the terminal could not be used for discharging.

The voltage between the exposed positive bus bar of module 14 and the battery case measured 25.4 volts, and about 3 volts were measured in other areas of the damaged battery.[46] Although the measured voltage was not as high as the dangerous value for DC voltage (50 to 60 volts), the values indicated the presence of energized components, and high-voltage risk was assumed. Options for cutting back or otherwise reducing the exposed bus bar were considered but could not be tried with the available tools. Instead, efforts were made to isolate the exposed terminal. The exposed orange high-voltage cables (not the cut loops mentioned earlier) were measured and were found not to have electrical potential.

The engineers tried to establish a computer connection to the battery management system under the SUV's rear seat. Access was accomplished by removing burnt material and cutting into the vehicle structure. The 12-volt battery powering the battery management system had been displaced in the

---

[46] An electrical bus, or bus bar, collects and distributes current, providing power to a vehicle's various subsystems.

crash. An alternate power source was established, and a computer was attached to the appropriate points. Attempts to communicate with the battery management system were unsuccessful. Because the battery could not be deenergized and information about the battery's status could not be obtained, the battery was inspected by hand, taking appropriate electrical and fire-safety measures.

The front of the SUV was extensively damaged. The high-voltage cut loop, normally located at the base of the windshield, was missing, as was the entire front electric motor and other components.[4746] The orange cables that led to the motor and other high-voltage connection points were exposed (figure 8). Attempts to measure the voltage were inconclusive, and any exposed terminals, as well as damaged battery components, were assumed to pose a high-voltage risk.

Figure 8. Damaged SUV with exposed orange high-voltage cables forward of front seat.

Removing the burnt materials at the front of the battery case, just forward of the front seats, revealed that the steel battery cover over modules 15 and 16 had peeled back and was positioned over modules 13 and 14 (refer to the battery diagram in figure 5, section 2.1.3). The modules behind modules 15

---

[47] Some electric vehicles, including the model X, have dual electric motors, one for the front wheels and another for the rear wheels.

and 16 appeared to be intact, although the case around modules 13 and 14 was damaged. The vents showed evidence of thermal activity at module 8 and at modules 10 through 14 (indicating that the battery's built-in thermal defenses worked as designed to discharge hot gases generated by thermal runaway).[48] Modules 15 and 16 were too damaged to establish the status of their vents. All other vents appeared to be undamaged.

The battery having reignited at least six times was evidence that it contained stranded energy, supported by the measurements of electric potential and the presence of intact cells. Removing the battery case for further inspection was deemed unsafe. Hence, the stranded energy remaining in the battery was not assessed.

2.3. Fort Lauderdale, Florida, May 2018

Source: CBS4 Miami.

Figure 9. Car burning next to residential driveway postcrash.

On Tuesday, May 8, 2018, at 6:46 p.m. eastern daylight time, a 2014 Tesla model S car, occupied by a driver and two passengers, was traveling on an urban road in Fort Lauderdale, Broward County, Florida. The car approached a curve while traveling at a recorded speed of 116 mph in a 30-mph zone.[49]

[48] As noted earlier, the valves in the vents were positioned to direct hot gases away from the passenger compartment.

[49] The curve had a posted advisory speed of 25 mph.

The car left the road and erupted in flames after it hit the wall beside a residential driveway. The car reentered the road, hit a light pole, and came to rest in the driveway of an adjacent residence (figure 9). The driver and front passenger suffered fatal injuries. The rear passenger sustained serious injuries.[50]

2.3.1. Initial Response

The Broward County call center received the first 911 call about the crash at 6:46 p.m. Police units arrived on scene at 6:49 p.m. Firefighters from Fort Lauderdale Fire Rescue were dispatched at 6:46 p.m. and arrived on scene 4 minutes later. When firefighters arrived, heavy flames were coming from the front of the car, and the heat was intense, according to the battalion chief on scene. The engine crew applied a foam/wetting agent to the fire. After the engine had used up half the 500 gallons of water in its tank, a supply line was established to a hydrant about 150 feet north of the engine, and firefighters continued attacking the fire.

After suppressing the fire in the car's interior, which took less than a minute, firefighters focused on the fire at the car's right front corner. Firefighters reported that the heat was intense and that they could see electrical arcing. The battalion chief told investigators that the fire came from a lithium-ion battery module that was lodged under the car's A-pillar.[51] The module had ruptured, and the individual battery cells inside were visible. Firefighters estimated using 200 to 300 gallons of water and foam to stop the flames and electrical arcing. Two large pieces of the battery had been completely separated from the vehicle and were found in the street. Although the pieces did not appear to be on fire, firefighters applied water and foam to them. The main part of the high-voltage battery was still connected to the car but had dropped to the ground.

The battalion chief told investigators that when he arrived on scene, he used the "Tesla app" (the company's online emergency response guide) to "get a detailed explanation of the vehicle, and where our cuts and no-cuts and things

[50] For further details, see the NTSB's report on its investigation of the Fort Lauderdale crash (NTSB 2019b).

[51] A car's *A-pillar* is an upright structural piece that supports the windshield and the front of the roof.

of that nature were. And it had indicated where the batteries were located up underneath the vehicle."[52]

2.3.2. Secondary Response

As part of the police investigation and cleanup, the vehicle and associated debris were loaded onto tow trucks. The operator of one truck told NTSB investigators that he had attended a Tesla training class and was familiar with the company's vehicles. He said that while he was winching the car onto his truck (about 7:45 p.m.), the entire battery case separated from the vehicle. The battery reignited, but a brief application of water and foam extinguished the fire. The operator said that the battery briefly reignited again when a chain passed over the battery case while the battery and two piles of debris were being loaded onto a separate truck (about 8:00 p.m.). The battery self-extinguished (fire suppression was not performed).

While being unloaded at the tow yard about 3 hours after the crash, the battery case and modules briefly arced and smoked, but the battery self-extinguished once again, without any fire suppression being performed. At the tow yard, the vehicle and debris were stored outside. The employees at the tow yard had previously worked with electric vehicles and were aware that they should be isolated. The employees put the piles of debris about 20 feet from other objects or vehicles. One pile consisted of the battery pack on a wooden pallet with other loose car parts, wrapped in clear plastic. The second pile consisted of debris wrapped in a heavy tarp.

2.3.3. Postcrash Inspection

On June 7, 2018, the NTSB met at the tow yard with representatives of the Fort Lauderdale Police Department and Florida Highway Patrol to inspect the vehicle (figure 10) and the debris piles. The items wrapped in the tarp were removed and placed near the vehicle. Individual loose battery cells were found in the tarp, as well as the right front tire and small debris. Large items, including the left front door, front electric motor, right front wheel assembly, and damaged battery modules that had separated from the vehicle (determined to be modules 15 and 16), were found on top of the battery case in the second debris pile (figure 11).

---

[52] A standard feature of the manufacturers' emergency response guides described in section 4.5 is a diagram showing where firefighters can safely cut the vehicle structure (to extricate the occupants) without hitting high-voltage elements or other hazards.

Figure 10. Car parked at tow yard showing severe crash and fire damage.

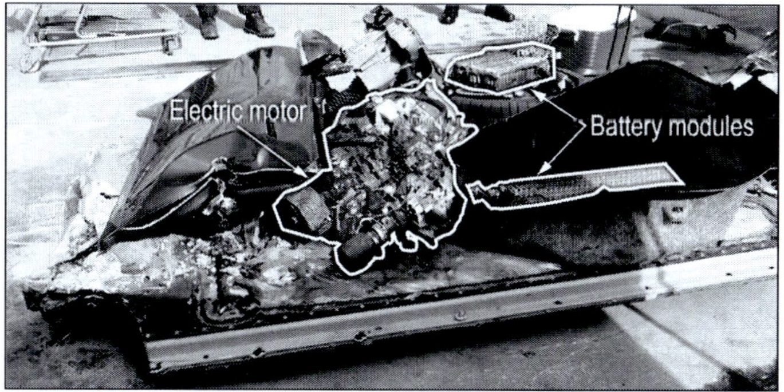

Figure 11. Debris pile on pallet with remains of electric motor and battery modules highlighted.

The high-voltage cut loop was found attached to the front electric motor and had not been cut. Firefighters told investigators that the cut loop was inaccessible during the fire response. Review of the cut loop, emergency response guidance, and exemplar vehicles revealed a minor inconsistency in the cut loop instructions, such that in some vehicles, the cut loop label might be obscured from view under a trim panel. This vehicle model, manufactured after June 2013, also had a cut loop at the left rear door frame, in case the front cut loop was not accessible. Reaching the rear cut loop required penetrating the door frame with a circular saw at a depth indicated by a label at the location of the cut loop. The label was found to have been attached incorrectly to the vehicle's right rear door, where no cut loop was installed. Firefighters had not attempted to use the rear cut loop.

The battery case was severely damaged and ruptured along the entire length of the front surface and extending into the area of modules 13 and 14, which were mounted just behind modules 15 and 16 (figure 12).[53] Two irregular holes had melted through the battery case, ranging from 3 to 5 inches in diameter. One was on the top of the battery case and would have been between the front seats in the intact vehicle. The other was on the bottom of the case and corresponded to a hole through the floor of the vehicle in roughly the same location. The photograph in figure 12 shows the steel cover on the top of the battery case. The location corresponds to the intersection of four battery modules (Nos. 11, 12, 13, and 14) and is the location of high-voltage connection points. The battery case was significantly warped and deformed in this area.

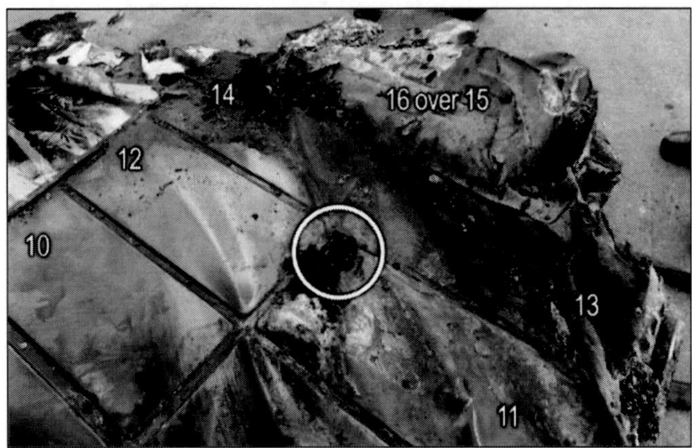

Figure 12. Top surface of forward part of battery case showing damage to front and hole melted through steel cover (circled), with modules numbered.

Investigators inspected the module vents and found either water and debris or the remnants of fire suppression foam in all venting channels. They found evidence of thermal damage and venting at the vents for modules 9, 12, 13, and 14.[54] The condition of the vents for modules 15 and 16 was obscured by crash damage. Investigators removed the cover of the battery pack to access

---

[53] As noted earlier, in all Tesla vehicles, the battery case is mounted under the floor.

[54] An intense fire had broken out near the car's front wheels, with loose battery cells shooting out of the case due to rapidly venting gas.

the high-voltage terminals and measured voltages across each module—at the forward terminal to ground, and at the aft terminal to ground. Module 14 had zero voltage. Modules 6 through 13 and modules 15 and 16 had voltages ranging from 3.3 to 46.9 volts. The voltages of modules 1 through 5 ranged from 69.9 to 167.3 volts. The measurements confirmed that stranded energy remained in the high-voltage lithium-ion battery.

2.4. West Hollywood, California, June 2018

On Friday, June 15, 2018, about 5:30 p.m. Pacific daylight time, a 2012 Tesla model S car was traveling on an urban road in West Hollywood, Los Angeles County, California. Other motorists saw smoke coming from the car and flagged down the driver. The driver stopped and left the vehicle, which then caught fire. The driver was uninjured. A Los Angeles Sheriff's Department patrol car stopped, and officers directed traffic around the burning car. After firefighters from a nearby Los Angeles County Fire Department station extinguished the fire, the vehicle was towed away without incident.[55]

Two people captured videos of the event—a bystander and the driver. Video taken by the bystander captured the left side of the car. Figure 13 is a still image taken from the bystander's video showing fires behind the left front wheel and in the left rear wheel well (large circles), as well as pieces of material ejected from the vehicle (small circles). The image was taken 26 seconds from the start of the bystander's recording and occurred within 2 minutes of the start of the incident.

Figure 13. Image from bystander's video with fire visible behind left front wheel and in left rear wheel well. (Source: Los Angeles County Sheriff's Department).

---

[55] Detailed information gathered in the NTSB's investigation of the West Hollywood battery fire can be found in the public docket (NTSB case number HWY18FH014).

2.4.1. Initial Response

According to the Los Angeles Fire Department's incident report, the alarm came in at 5:38 p.m., the fire engine arrived on scene at 5:40 p.m., and the last unit cleared the scene at 6:28 p.m. After firefighters extinguished the flames and cut the high-voltage disconnect loop, they applied water and foam to various locations, including the firewall and under the wheel wells.[56] Figure 14 is a still image from the video taken by the car's driver showing smoke coming from the front of the vehicle after firefighters had extinguished the flames. The image was taken 14 minutes 34 seconds into the recording.

Source: car's driver.

Figure 14. Image from driver's video showing car after flames were extinguished but with smoke still rising (circled).

The fire captain told NTSB investigators in a postincident interview that the flames were extinguished quickly, but that smoke continued to come from the vehicle. He was surprised that the vehicle continued to emit smoke and said that firefighters applied water for longer (about 30 minutes) than he would have expected for such an event. He estimated that firefighters used 300 gallons of water and foam and said that they did not connect to a hydrant.

In an effort to stop the smoke, firefighters removed parts of the vehicle's left front fender and hood trim, using a pry bar with a metal-cutting claw, and applied additional water and foam under the hood and behind the front wheels. The fire captain said that he was surprised that after the cut loop was severed, the vehicle's computer screen remained on. (The captain told investigators that he had been trained in electric vehicles and knew where to find the cut loop.)

[56] Model S cars manufactured before 2013 have only one cut loop, at the front.

After they had extinguished the flames, but with smoke still present, the owner called the vehicle manufacturer's representative. The fire captain spoke to the representative, who told him that the 12-volt battery that powered the computer was separate from the main battery, and that after the loop was cut, it would take 10 to 15 minutes for the 12-volt system to power down. The fire captain spoke to the representative about the continued smoke. He was told that firefighters needed to apply water to cool the high-voltage battery until it stopped smoking.

2.4.2. Secondary Response

The vehicle manufacturer's representative advised not to tow the car immediately, because of the risk that the battery could reignite. The fire captain decided, however, to move the vehicle from the scene in order to open the road to traffic. The car was taken to an impound lot. The manufacturer contacted the owner and received permission to inspect the vehicle; the company then bought the vehicle outright to do a complete investigation. The vehicle was taken from the impound lot to Burbank, California, for inspection at one of the manufacturer's service facilities (as described below). The car was deemed safe to transport. It was then taken to the manufacturer's research center in Sunnyvale, California, several hours north, for a complete inspection and battery teardown. The battery did not reignite at any time during the initial or secondary response.

2.4.3. Postincident Inspection

The NTSB was given access to photographs from the Burbank inspection. The battery pack was removed from the vehicle, and the cover of the pack was taken off so that the high-voltage terminals could be accessed. All modules were found to be intact except module 14, which was the only one to have experienced thermal runaway (refer to the battery diagram in figure 5, section 2.1.3). Module 14 was electrically isolated, and the other modules were discharged to about a 50 percent state-of-charge.[57] Reducing the battery's charge to 50 percent allowed for safe transport, and the vehicle manufacturer elected not to completely deenergize the battery so as to preserve evidence. The manufacturer informed the NTSB that its inspection had found no

[57] The vehicle manufacturer informed the NTSB that it deenergized the battery modules by connecting them to a proprietary positive temperature coefficient heater. Such heaters use conductive inks printed on polymer-based substrates to generate heat, rather than wires and coils, and are self-regulating, so they do not overheat.

indication that an impact or other external factor was involved in the battery's failure.

NTSB investigators attended a second inspection in Sunnyvale to view the damage to the vehicle, the battery, and the battery case and to observe the effects of thermal runaway. The vehicle inspection was followed by a teardown of the battery pack. The battery case exhibited holes at the left and mid-front that appeared to be caused by venting gas. Figure 15 shows the whole pack, with damaged areas circled and insets highlighting details of the damage. The damage created paths for vented gas and flames to reach both the left- and right-side overpressure vents of the battery pack. The damaged areas were photographed after debris and parts of the battery pack were removed.

3. Other High-Voltage Battery Fires

Figure 15. Battery pack with top cover removed to show location of holes in battery case and insets showing closeups of damaged areas.

Although all four investigations conducted by the NTSB (section 2) involved Tesla cars or SUVs, the issue under investigation was vehicle fires caused by high-voltage lithium-ion batteries, not vehicle design. In three of the four investigations, the batteries had been damaged in high-speed, high-severity crashes. In one case, the fire resulted from an internal battery failure. In all cases, the battery fires presented unique problems to first and second responders. As illustrated below, the high-voltage lithium-ion batteries in electric vehicles from other manufacturers have posed similar problems to emergency responders when the battery case or battery cells were damaged or failed internally.

3.1. Chevrolet Volt Fire After NCAP Test

In June 2011, a Chevrolet Volt caught fire at a test facility 3 weeks after a crash test performed on May 12. The crash test was carried out as part of NHTSA's New Car Assessment Program (NCAP) for 2011 model year vehicles and to verify compliance with the Federal Motor Vehicle Safety Standards (FMVSSs). NCAP is a federal consumer information program that evaluates the performance of new automobile designs against safety threats such as crashes.[58] (Other countries operate similar programs.)

The crash test performed on the Chevrolet Volt in May 2011 was a side-impact pole test (the car was crashed sideways into a rigid pole at a speed of up to 20 mph, with a dummy placed in the driver's seat). The car was rotated along its longitudinal axis (rolled) 90 degrees at the end of the test to check for fluid leaks. As a result of the test, NHTSA gave the Volt a five-star crash safety rating—the highest.

Three weeks after the test, the Volt caught fire while parked outside the test facility. The fire damaged four nearby vehicles. Forensic examination found that the transverse stiffener under the driver's seat had penetrated the Volt's battery compartment, damaged the lithium-ion battery, and ruptured the battery's liquid cooling system.[59] The damage was "not easily detected by visual inspection [and] went unnoticed at the time of the test," according to NHTSA's report on the incident (Smith 2012). Examiners determined that the fire was precipitated by damage to the battery cells and electric shorting. Coolant had leaked into the battery cells through the damaged battery case and dried out over time, creating short circuits from residual coolant salts.

In September 2011, NHTSA repeated the side-pole test on another Volt. The test did not replicate the results of the original—there was no intrusion

[58] NHTSA announced in a press release on October 16, 2019, that it planned to upgrade NCAP in 2020, the 40-year anniversary of the program.

[59] A *stiffener* is a structural support that strengthens and stabilizes the vehicle body. In this case, the stiffener was a metal beam that ran from side to side under the driver's seat.

into the battery compartment, no cell damage or shorting, no leaks, and no postimpact fire. The car was monitored for 3 weeks afterward, with no thermal or electrical activity observed in the vehicle or the battery. NHTSA then conducted impact tests on six Chevrolet lithium-ion batteries that had been removed from vehicles and attached to fixed supports. Three batteries caught fire, one as a result of another battery catching fire. NHTSA concluded that the fires were caused by shorting of the battery electronics, caused in turn by battery coolant that leaked into the battery compartment when the batteries were rolled over as part of the test.

In November 2011, NHTSA announced that it was opening a safety defect investigation to assess the risk of fire in Chevrolet Volts that had been involved in serious crashes.[60] At the conclusion of its investigation, NHTSA issued a press release stating, "NHTSA does not believe that Chevy Volts or other electric vehicles pose a greater risk of fire than gasoline-powered vehicles" and that the agency remained "unaware of any real-world crashes that have resulted in a battery-related fire involving the Chevy Volt or any other electric vehicle."[61] After the test incident, General Motors made design modifications to improve the battery case's resistance to crash forces.

3.2. International Examples

International cases of high-voltage lithium-ion battery fires have been documented. The NTSB was not involved in the investigations of any of those fires, and limited information is publicly available. Three cases that illustrate the issues such fires pose for emergency responders, and how emergency personnel responded, are described below.

3.2.1. Norway, March 2017

On March 30, 2017, the Norwegian Fire and Rescue Department (NFRD) in the region of Nedre Romerike used a donated, lightly damaged 2017 BMW i3 in a test and training exercise.[62] Participants in the exercise were the NFRD, the Norwegian Defense Research Establishment, the local police, a local ambulance service, an insurance company, and a battery disposal company.

---

[60] See press release dated November 25, 2011.

[61] See press release dated January 20, 2012.

[62] See letter from the NFRD dated September 18, 2018, in the NTSB public docket for this report.

The exercise consisted of cutting the vehicle's body and frame (without touching the battery), attempting to trigger a short circuit and fire by crushing the vehicle in different places with a bulldozer, and driving a large metal rod through the center of the vehicle to pierce the battery. Piercing the battery resulted in a fire, but firefighters could not stop the thermal runaway, and the fire was left to burn itself out.

Figure 16 illustrates the sequence of events using images from a video recording of the training exercise. About 2,000 gallons of water and foam were applied during the exercise.

3.2.2. Belgium, May 2017

On May 13, 2017, a Mitsubishi Outlander PHEV, equipped with a high-voltage lithium-ion battery, struck a tree in Dilsen-Stokkem, Belgium, and caught fire. According to an account published by the International Association of Fire and Rescue Services (CTIF), an emergency rescue team received a call that a car was against a tree with a person trapped inside.[63] While a fire engine and intervention team were traveling to the scene, another call informed the team that the vehicle might be on fire. The commander directed two firefighters to don breathing apparatus.

When firefighters arrived, smoke was coming from the car's motor compartment, and flames were visible. Firefighters brought the fire under control and extricated the driver by removing the car's B-pillar.[64] The driver was taken to a local hospital, where he died of his injuries.

Firefighters disconnected the car's 12-volt battery and applied water to cool the motor compartment. Before leaving the site, they checked the vehicle using a thermal imaging camera. While the car was being winched onto a truck, the fire reignited, preceded by a loud bang and a blue jet of flame, according to the tow truck driver. Responders put the car on its side, hoping to reach the high-voltage battery. After a "long time," they found the service plug, removed it while wearing the recommended PPE (electrically insulated gloves and face shields), and extinguished the fire. Firefighters accompanied the car to a safe place.

---

[63] See the account published by CTIF on March 7, 2018 (accessed November 12, 2020).

[64] The *B-pillar* is the structural vertical support behind a car's front door.

(A)

(B)

(C)

(D)

Source: NFRD video.

Figure 16. Series of images showing (a) attempt to trigger short circuit by crushing car with bulldozer; (b) flames erupting 8 minutes after piercing car with metal rod; (c) flames no longer visible, but battery continuing in thermal runaway; and (d) fire burning itself out, with arrow pointing to metal rod.

3.2.3. The Netherlands, March 2019

On March 25, 2019, in Tilburg, the Netherlands, firefighters dropped a BMW model i8 coupe into a water bath after the car began smoking while on display at a dealership (see figures 17 and 18).[65] The car was kept in the water for 24 hours. As the Central and West Brabant Fire Brigade explained in a message on its Facebook page, "Extinguishing an electric car requires a lot of water for a longer period of time, partly because the battery packs are difficult to reach and the fire flares up again cell after cell."[66]

---

[65] See various online news reports (autorevolution, March 26, 2019; Motor Illustrated, March 26, 2019; Motor 1 News, March 26, 2019; Energy Live News, April 4, 2019) and the BMW blog, March 26, 2019 (all accessed November 12, 2020). The incident is also described in Roman (2020) and is cited in a Swedish study on the safety of lithium-ion vehicle batteries (Bisschop and others 2019).

[66] (a) See original Facebook post, March 26, 2019 (Google translation) (accessed November 12, 2020). (b) The NTSB could not confirm that the car's high-voltage battery was the source of the fire.

Source: Central and West Brabant Fire Brigade.

Figure 17. BMW after firefighters attempted to extinguish fire, with crane positioned above and faint plume of smoke visible above car's windshield.

Source: Central and West Brabant Fire Brigade.

Figure 18. BMW being loaded into tank filled with water.

4. Regulatory and Industry Actions

Since the early 1990s, both national and international organizations have issued regulations and standards for electric vehicles and the high-voltage lithium-ion batteries that power them. Among the US organizations whose work relates to the safety issues faced by emergency responders are NHTSA, SAE International (SAE), and the NFPA. International entities include the United Nations Economic Commission for Europe (UNECE) and the International Organization for Standardization (ISO).

4.1. US Federal Motor Vehicle Safety Standard 305

To sell a vehicle in the United States, manufacturers must self-certify that it meets the FMVSS performance requirements. FMVSS 305 applies specifically to electric vehicles.[67] The standard was adopted in September 2000, in parallel with NHTSA's participation in the United Nations effort (described in the next section) to harmonize global standards for vehicle safety. The original purpose of FMVSS 305 was to reduce crash-related deaths and injuries that could occur because of electrolyte spilled from batteries, intrusion of batteries or electrical converters into the passenger compartment, or electric shock:

- Electrolyte is not permitted to spill into the passenger compartment, and no more than 5.0 liters of electrolyte are allowed to spill outside the vehicle within 30 minutes of a crash test and after a static rollover test (conducted after barrier crash tests; see below).
- Electric energy storage or conversion devices must be anchored to the vehicle and remain attached by at least one anchorage after crash tests. Devices positioned outside the occupant compartment must not intrude into the compartment.
- Each high-voltage source must meet one of three requirements: (1) it must be electrically isolated from the vehicle's chassis; (2) its voltage must be below levels considered safe from electric shock hazards (30

[67] FMVSS 305, codified at Title 49 *Code of Federal Regulations* (CFR) 571.305, was adopted in September 2000 and was effective on October 1, 2001 (65 *Federal Register* 57980). Its full title is "Electric-Powered Vehicles: Electrolyte Spillage and Electrical Shock Protection." The standard applies to passenger cars and to multipurpose passenger vehicles, trucks, and buses with a gross vehicle weight rating of ≤ 4,536 kilograms (10,000 pounds) that use electrical propulsion components whose working voltages are more than 60 volts DC (or 30 volts AC) and that can attain a speed greater than 40 kilometers per hour (km/hr; 25 mph) over a distance of 1.6 km (1 mile) on a paved level surface.

volts AC or 60 volts DC); or (3) it must be enclosed in a physical barrier to prevent direct human contact.[68]

The above requirements of FMVSS 305 are evaluated after an electric vehicle is crash-tested and then rolled on its longitudinal axis (rollover). The crash-test conditions are specified by FMVSS 208 (CFR 571.208, "Occupant Crash Protection"). The following tests are administered: (1) hitting a frontal barrier at up to 48 km/hr (30 mph); (2) being impacted from the side by a barrier moving at up to 54 km/hr (33.6 mph), and (3) being impacted from the rear by a barrier moving at up to 80 km/hr (50 mph).

FMVSS 305 calls for high-voltage batteries to be marked with a yellow triangle containing a jagged black arrow and bordered in black. The marking is not required for electrical isolation barriers that cannot be accessed without tools or for electrical connectors or the vehicle charge inlet. High-voltage cables and components must be identified with an orange covering. The exterior of a vehicle is not required to be marked to show the location of high-voltage cut loops or other means of disconnecting the high-voltage system. The standard does not specify a method for disconnecting the vehicle's high-voltage power.

FMVSS 305 has been amended nine times since it came into force in 2001. The most recent changes were made to bring US regulations into harmony with the electrical safety requirements of global technical regulations (GTRs) for hydrogen and fuel-cell vehicles (GTR 13) and electric vehicles (GTR 20).[69]

---

[68] The standard defines e*lectrical isolation* as "the electrical resistance between the high voltage source and any of the vehicle's electrical chassis divided by the working voltage of the high voltage source." Electrical resistance is measured in ohms. Minimum electrical isolation requirements for DC components and AC systems that have physical barrier protection (100 ohms/volt) are lower than for AC components that do not have a physical barrier (500 ohms/volt).

[69] FMVSS 305 was amended in 2017 to adopt the normal operation requirements of GTR 13 (82 *Federal Register* 44945) and in 2019 to harmonize with both GTR 13 and GTR 20 (84 *Federal Register* 6758). Work on GTR 13 began in 2005, with the purpose of establishing safety-related performance requirements for hydrogen-fueled vehicles that would achieve levels of safety equivalent to those for conventional internal combustion vehicles (a GTR working party had begun researching existing regulations and standards for alternative fueled vehicles in 1998). GTR 13 covers the safety of high-voltage electrical parts and includes requirements for electrical isolation to protect against electric shock, under both noncrash and postcrash conditions. GTR 13 was issued in 2013 (UNECE 2013; accessed November 12, 2020).

4.2. Global Technical Regulation for Electric Vehicles

Since 1952, the United Nations has led an effort to develop technical regulations for cars and other motor vehicles that can be coordinated, or harmonized, worldwide. An agreement signed in 1998, known as the 1998 Agreement, established a process for jointly developing GTRs related to motor vehicles, equipment, and parts.[70] The United States is a signatory (contracting party) to the 1998 Agreement. A global registry of harmonized regulations was created pursuant to the 1998 Agreement. Contracting parties use their countries' individual rulemaking processes to incorporate the GTRs into their national laws and regulations.

GTR 20, the first GTR focused on electric vehicle safety, was entered into the global registry on March 14, 2018 (UNECE 2018). GTR 20 is phase 1 of a process of establishing uniform global standards for electric vehicles. Phase 2 concerns thermal propagation tests, venting and management of gases released postcrash, and requirements for warning signals. A working group on electric vehicle safety, established by the World Forum for Harmonization of Vehicle Regulations (part of the UNECE), is developing the standards. NHTSA participates as the US delegation to the working group.

GTR 20 specifies both in-use requirements (covering normal vehicle use) and postcrash requirements. The in-use requirements address occupant safety for thermal events that can lead to fire, explosion, or smoke. Vehicles are required to provide advance warning of a hazardous situation inside the passenger compartment that will allow egress within 5 minutes, and manufacturers must make documents available describing the warning system (including a risk reduction analysis, diagrams, and engineering documents such as tests).

GTR 20 establishes four measures for determining the safety of electric vehicle occupants, rescue workers, and first responders after a crash. The requirements are to be met by a separate crash test. It is left to individual countries to specify the criteria according to their own crash-test requirements. At least one of the following measures to protect against postcrash electric shock must be met: (1) absence of high voltage (≤ 60 volts within 60 seconds of impact for high-voltage buses); (2) low electrical energy (the total energy

---

[70] The full title of the 1998 Agreement is "Agreement concerning the Establishing of Global Technical Regulations for Wheeled Vehicles, Equipment and Parts which can be fitted and/or used on Wheeled Vehicles." It came into force in 2000.

of impulse currents must be < 0.2 joules, from components to the vehicle chassis); (3) physical protection; and (4) isolation resistance.[71]

To protect against electric shock from indirect contact, resistance requirements are given for exposed conductive parts (between the parts and the electrical chassis or between simultaneously reachable conductive parts). As with the in-use requirements, the postcrash requirements for electrical isolation resistance specify minimum values between the high-voltage bus and the electrical chassis for powertrains consisting of separate or combined DC and AC buses.

Postcrash requirements for an electric vehicle's rechargeable electrical energy storage system include vehicle-based tests for electrolyte leakage, retention of the battery pack (it must remain attached to the vehicle and not intrude into the passenger compartment), and fire hazards (no evidence of fire or explosion for 1 hour after a crash test). The electrolyte leakage requirement allows no leakage from the battery into the passenger compartment and leakage of no more than 7 percent of the electrolyte (maximum 5.0 liters) outside the passenger compartment for 60 minutes after a crash. (No leakage is allowed if the electrolyte is not a liquid.)

4.3. SAE International Standards

SAE is a US-based professional society that creates standards, organizes technical meetings, and publishes technical papers through voluntary, collaborative efforts.[72] SAE recommended practice J2990 (*Hybrid and EV First and Second Responder Recommended Practice*) addresses the hazards faced by first and second responders to crashes and other incidents (such as garage fires) involving electric vehicles.[73] The recommended practice was first published in November 2012 and was reissued in July 2019. It emphasizes the chemical, electrical, and thermal risks associated with the high-voltage systems (including batteries) of electric vehicles and recommends best practices to "help protect emergency responders, tow and/or recovery, storage,

---

[71] The *joule* is the unit of work or energy in the International System of Units. One joule = 1 watt-second (the energy released in 1 second by a current of 1 ampere through a resistance of 1 ohm). One kWh = 3.6 million joules, and 1 joule = 2.77778×10^{-7} (0.000000277778) kWh.

[72] SAE originally focused on the American automotive industry but has since expanded globally and into other transportation industries, such as aerospace.

[73] Recommended practice J2990 can be purchased on the SAE website (accessed November 12, 2020).

repair, and salvage personnel." SAE J2990 considers only lithium-ion batteries and does not discuss the hazards associated with alternative fuels such as hydrogen (treated in another SAE publication).

4.3.1. Emergency Response Guides

SAE J2990 gives format and content recommendations for the emergency response guides produced by electric vehicle manufacturers. It recommends that manufacturers create quick reference sheets, following the guidance in ISO standard 17840: *Road Vehicles—Information for First and Second Responders* (see section 4.4 for details). It further recommends standardizing the emergency response guides using the ISO 17840 templates for organization and appearance (chapter headings, sequence of chapters, color codes, graphics) and for the design and color of fuel and energy labels.

SAE J2990 notes that manufacturers' guides "cover many of the necessary areas required for responders, but they vary in style and content . . . [and the] text is often written from an engineering and technical mindset, rather than from a responder's point of view." An appendix to SAE J2990 shows a 1-page quick reference guide (figure 19, from the NFPA emergency field guide [NFPA 2018]) that summarizes critical information about fire suppression in hybrid and electric vehicles.

SAE J2990 recommends that emergency response guides "should contain crucial and in- depth information linked to the quick reference sheet" and lists examples of available responder guides, including the NFPA emergency field guide, and how they can be obtained (individual manufacturer websites, commercial platforms that must be purchased, and smartphone applications).[74]

[74] See section 4.5 for information about the NFPA emergency field guide and the emergency response guides from individual manufacturers.

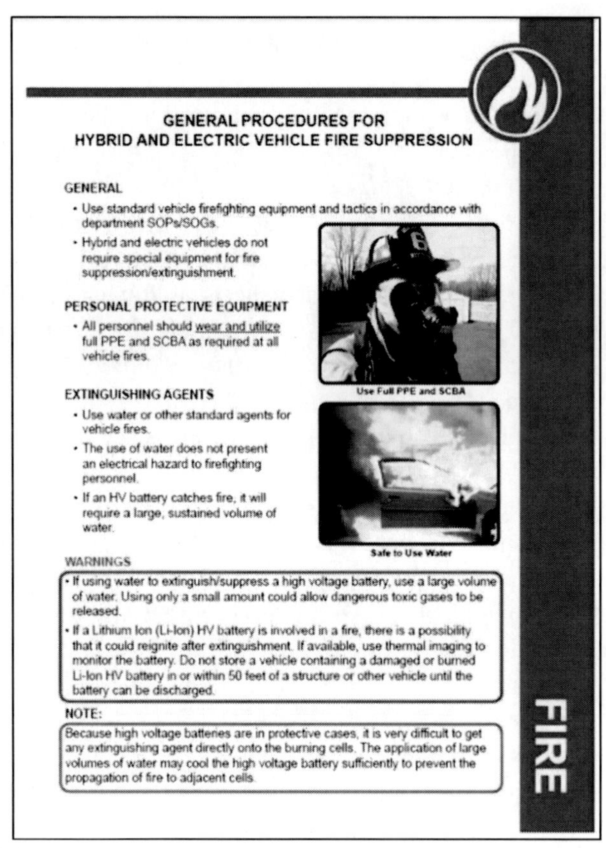

GENERAL PROCEDURES FOR
HYBRID AND ELECTRIC VEHICLE FIRE SUPPRESSION

GENERAL

- Use standard vehicle firefighting equipment and tactics in accordance with department SOPs/SOGs.
- Hybrid and electric vehicles do not require special equipment for fire suppression/extinguishment.

PERSONAL PROTECTIVE EQUIPMENT

- All personnel should wear and utilize full PPE and SCBA as required at all vehicle fires.

Use Full PPE and SCBA

EXTINGUISHING AGENTS

- Use water or other standard agents for vehicle fires.
- The use of water does not present an electrical hazard to firefighting personnel.
- If an HV battery catches fire, it will require a large, sustained volume of water.

Safe to Use Water

WARNINGS

- If using water to extinguish/suppress a high voltage battery, use a large volume of water. Using only a small amount could allow dangerous toxic gases to be released.
- If a Lithium Ion (Li-Ion) HV battery is involved in a fire, there is a possibility that it could reignite after extinguishment. If available, use thermal imaging to monitor the battery. Do not store a vehicle containing a damaged or burned Li-Ion HV battery in or within 50 feet of a structure or other vehicle until the battery can be discharged.

NOTE:

Because high voltage batteries are in protective cases, it is very difficult to get any extinguishing agent directly onto the burning cells. The application of large volumes of water may cool the high voltage battery sufficiently to prevent the propagation of fire to adjacent cells.

FIRE

Source: NFPA.

Figure 19. One-page emergency response guide ("quick reference sheet") for fire suppression in electric vehicles. SOPs/SOGs = standard operating procedures/standard operating guidelines. HV = high voltage.

4.3.2. Disabling High-Voltage Systems

SAE J2990 recommends three methods for disabling (disconnecting) high-voltage systems, noting that the methods "generally do not completely remove high voltage from the vehicle" but are intended "to limit its distribution around a vehicle": (1) automatic shutdown; (2) switching ignition switch to OFF (which should disconnect the high-voltage system from the high-voltage sources and discharge the system to ≤ 60 volts DC or 30 volts AC within 10 minutes); (3) cut or disconnect battery cables to discharge the 12-volt system, and cut or disconnect the 12-volt output cable. SAE J2990 further recommends that at least two methods be incorporated into the design of an electric vehicle.

SAE J2990 states that removing an electric vehicle's manual disconnect should not be a primary method for first responders to disable the high-voltage circuits, because (1) the variety of designs makes locating and activating manual disconnects inefficient, (2) first responders do not always have the required PPE, and (3) the manual disconnect might be inaccessible.[75] The recommended practice includes design considerations for manual disconnects if manufacturers prefer removing them as the method for disabling high-voltage systems.

4.3.3. Postincident Vehicle Inspection

SAE J2990 recommends two inspection stages after a crash or other incident to make certain that the high-voltage system has shut down and is not damaged—one at the incident scene and one at the storage site afterward. The vehicle should remain physically isolated until it has passed inspection. Recommended inspection steps, and what actions to take, are listed in table 1.

Table 1. Postincident inspection steps recommended by SAE J2990

Step	Action	Notes
1	Inspect for signs of fire or smoldering.	Use thermal camera or infrared temperature probe if possible.
2	Listen for gurgling, bubbling, crackling, hissing, or popping noises from battery.	Sounds can indicate venting of overheated cells or arcing in high-voltage system.
3	If groups of battery cells have separated from battery enclosure, alert responders of potential exposure to high voltage or fire reignition.	Contact equipment manufacturer for depowering recommendations, packaging instructions, and disposal recommendations. If sufficient information is not available, consult latest version of US Department of Transportation/Transport Canada Emergency Response Guidebook for lithium-ion batteries (guide 147) or NiMH (guide 171).[a]

[75] *Manual disconnects*, also called *manual service disconnects*, are devices such as plugs, levers, or switches that emergency responders can manipulate to disconnect an electric vehicle's high-voltage system. The devices are found in various locations—for example, behind the back seat or near a rear tire—depending on the vehicle make and model.

Table 1. (Continued)

4	If vehicle is submerged, do not remove submerged service disconnect, but turn off ignition if possible. Disable vehicle by chocking wheels, placing in park, and removing ignition key or disconnecting 12-volt battery.	Understand that electric vehicles are designed to be safe in water. Small bubbles emanating from vehicle do not create shock hazard. Water damage to electrical components could lead to reignition. Do not store vehicle that has been submerged indoors until high-voltage energy is depowered.
5	Ensure that high-voltage system is disabled.	Refer to manufacturer's emergency response guide or emergency field guide to verify. At a minimum, disable 12-volt system.
6	Examine mechanical integrity of battery system.	Is enclosure ruptured, cracked, punctured, or dented?
7	Inspect for evidence of fire or heat damage.	Signs include smoke residue or heat damage around battery system and burnt odor from battery system.
8	Inspect for evidence of arcing in high-voltage system. Notify tow truck driver of potential hazards and recommendations for isolation.	Carbon traces indicate that isolation of high-voltage system has been lost.
9	Inspect for evidence of external battery leaks. Notify tow truck driver of potential hazards and isolation requirements.	Electrolyte of lithium-ion battery has sweet odor, like ether, that could indicate battery leak. Leaking electrolyte normally creates drops, not puddles.

[a] A new edition was published in August 2020. The guidebook is intended for use by firefighters, police, and other emergency personnel who may be first to arrive at the scene of a transportation incident involving dangerous goods. Available online in English and Spanish.

SAE J2990 recommends towing a damaged electric vehicle on a flatbed, to avoid generating voltage from the turning wheels. If the vehicle's wheels must be turned—because it has run off the road, for example—its speed should be kept below 5 mph. After being loaded onto a tow truck, the vehicle's structural integrity should be checked. If the vehicle rolls while it is on the tow truck, the inspection steps listed above should be repeated.

SAE J2990 states that tow operators should arrange to tow the vehicle to an offsite location where it can be isolated. Once there, the vehicle should be inspected again. It should also be inspected for evidence of internal battery leaks, which could lead to short circuits or loss of high-voltage isolation, and the battery should be examined for loss of mechanical integrity. If airbags have deployed, further diagnostic steps should be conducted to assess the integrity of the high-voltage system, such as measuring the battery temperature.

SAE J2990 recommends two barrier methods for an electric vehicle during storage: (1) separate the vehicle from combustibles and structures by 50 feet on all sides, or (2) create a barrier of earth, steel, concrete, or solid masonry around the vehicle.

4.3.4. Hazard Communication

SAE J2990 states that a 24-hour telephone hotline provided by the vehicle manufacturer is "not an appropriate communications medium for first and second responders at this time" (for one thing, subject matter experts are widely dispersed and making them constantly available is not practical). Instead, it recommends that manufacturers should make emergency response guides available in digital format at any time, accessible through links from a website, and that the information should be made readily available to third parties.

4.4. ISO Standard 17840

The ISO is a worldwide federation of national standards bodies. It publishes international standards on thousands of topics, including quality management, environmental management, health and safety, energy, food safety, and information technology. The work is carried out by technical committees. Any member interested in a particular subject can be represented on the corresponding technical committee. The United States is a member body.

ISO standard 17840 (*Road Vehicles—Information for First and Second Responders*) is a set of four documents defining the structure and content of the information that vehicle manufacturers provide for emergency responders to vehicle fires or crashes.[76] The standard was created by CTIF, with the collaboration of the European New Car Assessment Programme (Euro NCAP) and Europe's Schengen Information System.[77]

ISO 17840-1 (published in August 2015) standardizes the content and layout of rescue sheets for passenger cars and light commercial vehicles. Rescue sheets are short documents that give quick information about a vehicle's construction, intended for use by emergency responders at the scene of a crash. The standard covers conventional (diesel, gasoline), liquefied petroleum gas, compressed natural gas, electric, and hybrid electric vehicles. ISO 17840-2 (published in April 2019) standardizes the rescue sheets for buses, coaches, and heavy commercial vehicles.

ISO 17840-3 (published in April 2019) establishes a template and defines the general content for manufacturers' emergency response guides—longer documents that give in-depth "necessary and useful information" about a vehicle involved in an incident. Emergency response guides are intended for use in conjunction with rescue sheets, generally for training emergency responders on unconventional vehicles. This part of ISO 17840 applies to passenger cars, buses, coaches, and light and heavy commercial vehicles.

ISO 17840-4 (published in June 2018) defines the labels and colors used to indicate the fuel or energy used to propel a vehicle. For example, an orange diamond with a white zigzag in the center indicates high voltage. Use of the labels includes the rescue sheets defined in ISO 17840-1 and -2 and the emergency response guides covered in ISO 17840-3.

The text introducing ISO 17840-3 ties the standard to the quick decision-making required of emergency responders at a crash site, and to the need for emergency responders to avoid risking their own lives while saving others. The template established by the standard is described as "a flowchart for the main actions of first and second responders arriving at an accident scene" and as a means of promoting "the correct action with respect to the vehicle technology concerned." The template gives headings for the various sections of an emergency response guide (such as "Disable direct hazards/safety regulations" or "In case of fire") and the order in which they should appear (see figure 20) but leaves it to manufacturers to fill in specific information for each section. Standards for colors and graphics are given, including pictograms for indicating where tanks, batteries, and other components are

---

[76] The standards are available for purchase online from the ISO or from the ANSI webstore (both accessed November 12, 2020).

[77] The NFPA is the US representative organization to CTIF.

located in a vehicle. The headings and sequence of sections are the same for both rescue sheets and emergency response guides.

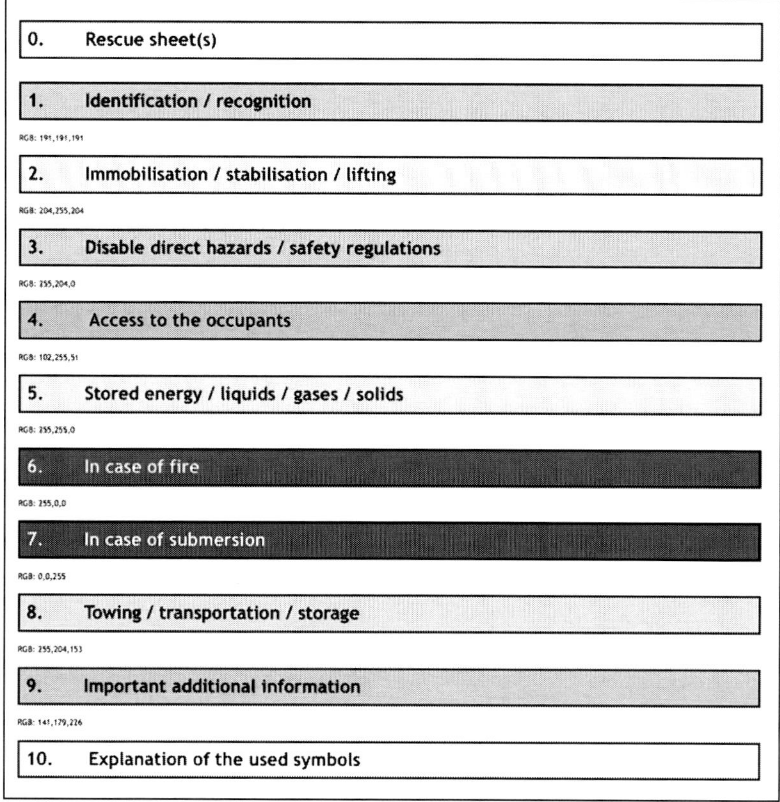

(Source: CTIF).

Figure 20. Contents of vehicle rescue sheets and emergency response guides as defined in ISO standard 17840.

In February 2020, Euro NCAP published a protocol for rescue, extrication, and safety testing and assessment (version 1.1), which promotes the use of ISO standard 17840 and incorporates scoring relative to the availability of a manufacturer's rescue sheets.[78] The protocol was prepared in coordination with CTIF. It promotes the availability of rescue sheets because "rescue services require detailed but readily-understood information regarding

[78] The protocol is described at www.euroncap.com (accessed November 12, 2020). Points are awarded for using the ISO standard.

the construction of individual vehicles to extract trapped occupants as quickly and safely as possible," which "is becoming more pressing as vehicles use different sources of power (e.g. electric/hybrid, hydrogen)."[79] The Australasian NCAP, which publishes crash test results for vehicles sold in Australia and New Zealand, published version 1.1.1 of a similar protocol in July 2020.[80]

Euro NCAP announced in June 2020 that it had centralized the manufacturers' rescue sheets in a free, downloadable mobile application for emergency responders, called "Euro Rescue." Australasian NCAP launched its own application, "ANCAP Safety," at the same time. In its press release, Euro NCAP stated: "As vehicles have become tougher, more complex and alternatively powered, it has become increasingly crucial that first responders know what they can and can't do at the scene of an accident."[81]

4.5. Industry Guidance for Emergency Responders

The 40-plus manufacturers of electric or other alternatively fueled vehicles voluntarily produce guidelines for firefighters and other emergency personnel who respond to incidents involving their vehicles. Links to the manufacturers' emergency response guides are available on the NFPA website, as well as on individual manufacturers' websites.[82] The NFPA also publishes a 460-page emergency field guide (NFPA 2018) that can be ordered and downloaded from its website. The NFPA recently added to its website a training video for firefighters, titled "Stranded Energy—How Little You Know Might Shock You."[83] A video prepared for the 2019 North American Heavy Rescue Symposium by the NFPA and others demonstrates best practices for responding to electric vehicle fires.[84]

---

[79] Quotations are from the protocol description at www.euroncap.com (accessed November 12, 2020).

[80] See the protocol document (accessed August 11, 2020).

[81] See press release, June 17, 2020 (accessed November 12, 2020).

[82] Smartphone applications and commercial platforms, such as Moditech, are also available. Some are free, others must be purchased. The Boron Extrication company website has links similar to the NFPA's (accessed November 12, 2020).

[83] The video can be viewed on the training and certification section of the NFPA website (accessed June 5, 2020).

[84] The video demonstrates using a thermal imagery camera to check the temperature of the battery and propping up a vehicle to apply water directly to the battery case. The video was sponsored by the NFPA, the California Department of Forestry and Fire Protection, and the training company Advanced Extrication, with the support of Tesla (accessed June 8, 2020).

4.5.1. NFPA Emergency Field Guide

The NFPA is a nonprofit organization that works to improve fire safety and provides training and education. The organization's emergency field guide is intended as a quick reference to use on scene or as a study aid. It begins by laying out general procedures for the initial response to vehicle crashes, fires, submersion in water, spills, and leak hazards. A separate chapter gives guidance for first aid in case of exposure to battery electrolytes or to natural gas or propane. The bulk of the field guide is devoted to brief, vehicle-specific entries drawn from the manufacturers' emergency response guides but not intended to replace them.

4.5.1.1. General Procedures

Initial response. Procedures are given for identifying the type of vehicle and its fuel and for immobilizing and stabilizing the vehicle. The guide states, "Always assume the vehicle is some type of hybrid, electric or alternatively fueled vehicle until proven otherwise." To immobilize a vehicle, emergency responders should approach it at a 45-degree angle, to stay out of the way in case it moves. The first step in stabilizing a vehicle is to turn off the ignition and disconnect the 12-volt battery. An alternative shutdown method is given if responders cannot access the ignition.

Extrication. Steps are given to extricate occupants from inside a crashed vehicle. Separate procedures are given for hybrid and electric vehicles, hydrogen fuel cell vehicles, and gaseous fuel vehicles. All vehicles must be immobilized and stabilized first. For hybrid and electric vehicles, responders should visually check for exposed high-voltage components or cables. Responders are warned to always assume that the high-voltage system is energized. Warnings are also given not to cut orange high-voltage cables (different from the 12-volt battery cables and the low-voltage cut loops that disconnect the high-voltage circuits) and not to penetrate high-voltage components. Further, firefighters are warned that if the 12-volt system cannot be disabled, occupant protection systems, such as airbags, could remain active, even if the high-voltage system is shut down.

The extrication section outlines procedures for handling damaged high-voltage batteries.

The guidance—

- Stresses the need for PPE and SCBA.

- Warns of leaking fluids, sparks, smoke, or bubbling noises from the high-voltage battery and instructs responders to vent the vehicle to avoid a buildup of fumes.
- Warns that sparks, smoke, and bubbling noises are signs of a potentially overheating battery, which could result in a delayed fire.
- Warns of a significant shock hazard (and to avoid contact with the battery).

A section on manual service disconnects notes that removing a manual disconnect will not discharge the high-voltage battery.

Fire extinguishment. The general procedures for extinguishing fires in alternative fuel vehicles state that "all personnel should wear and utilize full PPE and SCBA as required at all vehicle fires." Procedures for hybrid and all-electric vehicles include using standard firefighting equipment and tactics; no special equipment is required for extinguishing electric vehicle fires. The difficulty of extinguishing a high-voltage battery fire depends on (1) the size and location of the battery, (2) the extent of fire in the battery, (3) access to the battery and the "ability of the extinguishing agent to be applied to the battery assembly case," and (4) whether openings in the battery case allow the extinguishing agent to be applied directly to the burning cells. Points made about extinguishing agents are as follows:

- Use water or other standard agents for electric vehicle fires.
- Water does not pose an electrical hazard to firefighters.
- If a high-voltage battery catches fire, "it will require a large, sustained volume of water" (2,600 gallons or more, depending on the battery's size and location). Responders should establish a sustained water supply (hydrant or static water source).[85]

The following warnings are emphasized:

- If using only water to extinguish or suppress a high-voltage battery fire, use a large amount. A small amount of water could release toxic gases.
- A fire in a lithium-ion battery could reignite. The following steps should be taken:

[85] *Static water source* means a lake, stream, or pond.

- Use thermal imaging if available to monitor the battery (to determine the presence of fire).
- Do not store a vehicle with a damaged or burning lithium-ion battery within 50 feet of a structure or another vehicle until the battery is discharged.
- Reignition is accompanied by a "whooshing" or "popping" sound, followed by off-gassing of white smoke or electrical arcs or sparks that reignite with visible flames or fire. Reignition can occur within several hours to a day or more after the fire is extinguished.

The guidance notes the following about fires in high-voltage batteries:

- It is difficult to direct an extinguishing agent onto burning cells because high-voltage batteries are in a protective case. Applying a large volume of water might cool the battery enough to prevent the fire from propagating to adjacent cells. Applying water to a localized area for a long time leads to quicker extinguishment. Water should be applied even after no flame is visible.
- Anticipate longer fire-suppression times if the high-voltage battery is involved. Tests show that suppression can take an hour or longer.

4.5.1.2. Vehicle-Specific Guidance

Nearly 400 pages of the NFPA emergency field guide give specific initial response procedures for 43 makes and 195 models (counting generations of the same model separately) of alternatively fueled vehicles, including trucks, buses, and vans as well as cars. The guidance was prepared by the manufacturers and gathered together in the emergency field guide. The entries are organized alphabetically by manufacturer and then by vehicle model. Each entry is presented in a 2- or 3-page format.

4.5.2. Manufacturers' Emergency Response Guides

As stated earlier, links to the emergency response guides published by the manufacturers of alternative fuel vehicles can be found on the NFPA website. The downloadable guides can be accessed by clicking on "Alternative Fuel Vehicles Safety Training" at the bottom of the NFPA webpage, then clicking on "Emergency Response Guides." Altogether, 41 manufacturers are listed.[86] The guides include quick response sheets (2 to 4 pages), which are similar to the vehicle-specific guidance published in the NFPA emergency field guide but not necessarily identical. Some manufacturers also publish longer emergency response guides (such as BMW's 108-page rescue manual), which are included in the links. The guides are organized by model and year and cover hybrid electric vehicles, BEVs, fuel-cell vehicles, and vehicles powered by compressed natural gas. Seven manufacturers of electric buses, commercial trucks, or vans are included.

The guides generally start with information about how to identify a vehicle, including markings and vehicle identification numbers, then go on to describe various aspects of vehicle operation. The location of high-voltage components is usually shown on diagrams or photographs. Some diagrams label the parts; others rely on color coding and a key printed under the diagram. Information about airbags is often given, including their location and the safety importance of deactivating the 12-volt system that powers the airbags (to prevent the airbags from triggering during an emergency operation). The organization and type of hazard information varies.

The NTSB reviewed the manufacturers' response guides and assessed whether they covered 11 categories of information deemed critical to emergency personnel who respond to high-voltage battery fires. Only the subset of 36 manufacturers whose vehicles use high-voltage lithium-ion batteries was considered. What constituted critical information was determined from the NTSB's review of the regulations and standards related to the risks faced by emergency responders (see sections 4.1 to 4.4), as well as from its investigation of the emergency response to four electric vehicle fires (see section 2).

Table 2 lists the criteria the NTSB used to determine whether (yes or no) an emergency response guide covered a particular item of critical information for first and second responders. Determining whether the response guides included elements found in the NFPA emergency field guide and its vehicle-specific guidance was part of the review. Table 3 gives the results of the review.

---

[86] The downloadable guides include two manufacturers (Karma and BYD) that are not listed in the NFPA emergency field guide. The NFPA's emergency field guide contains vehicle-specific guidance from four manufacturers that are not on the downloadable list (Electric Vehicles International, Ferrari, Smart, and XLhybrids).

Table 2. Criteria for yes/no determination of critical information in manufacturers' emergency response guides, with corresponding column of table 3

Criteria for Yes/No Determination	Table 3 Column
Does format of guide conform to ISO standard 17840? Format follows organizational structure and order of information in ISO 17840.	2
Does guide contain elements in NFPA guide for that vehicle? Elements in NFPA guide include identifying vehicle, immobilizing vehicle, disabling vehicle, warnings (high voltage, airbag, silent movement), and extricating occupants.	3
Does guide contain vehicle-specific instructions for disconnecting high-voltage battery from rest of vehicle? Examples include use cut loops, remove plugs, pull levers, remove fuses, disconnect 12-volt battery.	4
Does guide include information on vehicle stabilization and use of PPE? Stabilization examples include placing vehicle in park or chocking wheels. PPE examples include wearing SCBA for fighting vehicle fires and high-voltage protective gear when handling high-voltage components.	5
Does guide include general information about high-voltage battery damage or fire? Examples include warnings that (1) damaged batteries can vent toxic gas, (2) high-voltage battery fires require copious amounts of water, and (3) high-voltage batteries can ignite after being extinguished.	6
Does guide contain vehicle-specific information about fighting high-voltage battery fire? Examples include (1) specific locations to apply water and (2) procedures for applying water to extinguish fire and cool battery. Some guides state that occupant compartment must be vented, or that fire or heat can ignite components near battery.	7
Does guide contain information about submerged vehicle? Examples include statement that vehicle does not pose risk of electric shock, or procedures to mitigate risk.	8
Does guide contain information about postincident recovery or towing of damaged vehicle? Examples include that flatbed should be used for transport and vehicle should be monitored for reignition.	9
Does guide contain information about storing damaged electric vehicle? Examples include that vehicle should be isolated outdoors, 50 feet from other fire sources.	10
Does guide contain specific information about mitigating risks of stranded energy? Examples include procedures for minimizing risks of battery reignition; also information for contacting manufacturer.	11
Does guide contain information about vehicle's high-voltage battery? Examples include peak voltage of battery pack or battery storage capacity.	12

Table 3. Presence or absence of critical information in emergency response guides for 36 electric vehicles equipped with high-voltage lithium-ion batteries

1 Vehicle make and model	2 ISO 17840 format	3 Elements of NFPA vehicle-specific guides	4 Vehicle-specific instruct-tions for HV disconnect	5 Stabiliza-tion and PPE information	6 General informa-tion on HV battery damage or fire	7 Vehicle-specific instruct-tions for HV battery fire	8 Submer-ged vehicle informa-tion	9 Damaged EV recovery/ towing informa-tion	10 Damaged EV storage informa-tion	11 Informa-tion on mitigating stranded energy risk	12 HV battery characteristics	13 Notes
Acura MDX Sport Hybrid EV	No	Yes	Yes	No	No	No	Yes	Yes	Yes	No	Yes	--
Audi A8 TFSI e	No	Yes	Yes	No	No	No	No	No	No	No	No	--
Azure Transit Connect Van EV[a]	No	Yes	Yes	No	Yes	No	No	Yes	No	No	Yes	--
BMW i3 EV 2015– 2018 and BMW Multi-Vehicle 2017	No	Yes	Yes	Yes	Yes	No	Yes	Yes	Yes	No	Yes	--

1 Vehicle make and model	2 ISO 17840 format	3 Elements of NFPA vehicle-specific guides	4 Vehicle-specific instruct-tions for HV disconnect	5 Stabiliza-tion and PPE information	6 General informa-tion on HV battery damage or fire	7 Vehicle-specific instruct-tions for HV battery fire	8 Submer-ged vehicle informa-tion	9 Damaged EV recovery/ towing informa-tion	10 Damaged EV storage informa-tion	11 Informa-tion on mitigating stranded energy risk	12 HV battery characteristics	13 Notes
Buick LaCrosse and Regal Hybrid EV 2012–2016	No	Yes	Yes	No	No	No	Yes	No	No	No	Yes	--
BYD K9M Series Bus EV	No	Yes	Yes	Yes	Yes	No	No	No	No	No	Yes	Vehicle has automatic fire suppression system.
Cadillac ELR Extended-Range EV 2014–2016	No	No	Yes	No	Yes	No	Yes	No	No	No	Yes	--
Chevrolet Spark EV 2014–2016	No	No	Yes	No	Yes	No	Yes	No	No	No	Yes	--
Chrysler 2017 Pacifica Hybrid	No	Yes	Yes	Yes	Yes	No	Yes	Yes	Yes	No	Yes	--

Table 3. (Continued)

1 Vehicle make and model	2 ISO 17840 format	3 Elements of NFPA vehicle-specific guides	4 Vehicle-specific instruct-tions for HV disconnect	5 Stabiliza-tion and PPE information	6 General informa-tion on HV battery damage or fire	7 Vehicle-specific instruct-tions for HV battery fire	8 Submer-ged vehicle informa-tion	9 Damaged EV recovery/ towing informa-tion	10 Damaged EV storage informa-tion	11 Informa-tion on mitigating stranded energy risk	12 HV battery characteristics	13 Notes
Fiat 500e EV 2013–2018	No	Yes	Yes	Yes	Yes	No	Yes	No	No	No	Yes	--
Fisker Karma PHEV 2011–2012[a]	No	Yes	Yes	Yes	Yes	No	Yes	Yes	No	No	No	--
Ford 2018 Focus EV	No	Yes	Yes	Yes	Yes	No	Yes	Yes	Yes	No	Yes	Guide gives manufacturer contact information.
Gillig Bus EV 2016–2018	No	No	Yes	No	Yes	No	No	No	No	No	Yes	--
GMC Sierra Hybrid EV 2016–2018	No	No	Yes	No	Yes	No	Yes	No	No	No	Yes	--

1 Vehicle make and model	2 ISO 17840 format	3 Elements of NFPA vehicle-specific guides	4 Vehicle-specific instruct-tions for HV disconnect	5 Stabiliza-tion and PPE information	6 General informa-tion on HV battery damage or fire	7 Vehicle-specific instruct-tions for HV battery fire	8 Submer-ged vehicle informa-tion	9 Damaged EV recovery/ towing informa-tion	10 Damaged EV storage informa-tion	11 Informa-tion on mitigating stranded energy risk	12 HV battery characteristics	13 Notes
Honda Clarity EV 2017–2018	No	Yes	Yes	Yes	Yes	No	Yes	Yes	Yes	No	Yes	--
Hyundai Ioniq EV 2017–2018	No	Yes	Yes	Yes	Yes	No	Yes	Yes	Yes	No	Yes	Guide instructs firefighters to monitor battery with thermal camera before leaving scene.
Infiniti Q50 Hybrid 2014–2018	No	Yes	Yes	Yes	Yes	N	Yes	No	No	No	Yes	--
Karma Revero 2017–2018	No	Yes	Yes	Yes	Yes	No	Yes	Yes	No	No	Yes	--
Kia Soul EV 2015–2018	No	Yes	Yes	Yes	Yes	No	Yes	Yes	No	No	Yes	--
Lexus LC500hl 2018	No	Yes	Yes	Yes	Yes	No	Yes	Yes	Yes	No	Yes	--

Table 3. (Continued)

1 Vehicle make and model	2 ISO 17840 format	3 Elements of NFPA vehicle-specific guides	4 Vehicle-specific instruct-tions for HV disconnect	5 Stabiliza-tion and PPE information	6 General informa-tion on HV battery damage or fire	7 Vehicle-specific instruct-tions for HV battery fire	8 Submer-ged vehicle informa-tion	9 Damaged EV recovery/ towing informa-tion	10 Damaged EV storage informa-tion	11 Informa-tion on mitigating stranded energy risk	12 HV battery characteristics	13 Notes
Lincoln MKZ Hybrid 2013–2018	No	Yes	Yes	Yes	Yes	No	Yes	Yes	Yes	No	Yes	Guide gives manufacturer contact information.
Mercedes-Benz	No	Yes	Yes	Yes	Yes	No	Yes	Yes	Yes	No	Yes	Guide contains 1-page rescue cards for all Mercedes-Benz passenger cars.
MINI Countryman PHEV 2017–2018	No	Yes	Yes	Yes	Yes	No	Yes	Yes	Yes	No	Yes	--
Mitsubishi Outlander PHEV 2018	No	Yes	Yes	Yes	Yes	Yes	Yes	Yes	Yes	No	Yes	Guide instructs to flood vehicle with water.
Nissan Leaf EV 2018	No	Yes	Yes	Yes	Yes	No	Yes	Yes	Yes	No	Yes	--

1 Vehicle make and model	2 ISO 17840 format	3 Elements of NFPA vehicle-specific guides	4 Vehicle-specific instructions for HV disconnect	5 Stabilization and PPE information	6 General information on HV battery damage or fire	7 Vehicle-specific instructions for HV battery fire	8 Submerged vehicle information	9 Damaged EV recovery/towing information	10 Damaged EV storage information	11 Information on mitigating stranded energy risk	12 HV battery characteristics	13 Notes
Nova Bus LFSe EV 2016–2019	No	Yes	Yes	Yes	No	No	No	No	No	No	Yes	--
Porsche AG Taycan Data Sheets 2019	No	Yes	Yes	Yes	Yes	No	Yes	Yes	Yes	No	Yes	Guide gives source for additional information.
Proterra Catalyst E2 EV Bus 2017–2918	No	Yes	Yes	Yes	Yes	Yes	No	No	No	No	Yes	Guide says to focus water at vents, use thermal camera until all heat sources are extinguished.
Scion iQ EV 2013[b]	No	Yes	Yes	Yes	Yes	No	Yes	Yes	Yes	No	Yes	Guide gives advice for offensive fire attack (flood battery with water) or defensive attack (let it burn out).

Table 3. (Continued)

1 Vehicle make and model	2 ISO 17840 format	3 Elements of NFPA vehicle-specific guides	4 Vehicle-specific instruct-tions for HV disconnect	5 Stabiliza-tion and PPE information	6 General informa-tion on HV battery damage or fire	7 Vehicle-specific instruct-tions for HV battery fire	8 Submer-ged vehicle informa-tion	9 Damaged EV recovery/ towing informa-tion	10 Damaged EV storage informa-tion	11 Informa-tion on mitigating stranded energy risk	12 HV battery characteristics	13 Notes
Smith Newton Step Van 2012–2017a	No	No	Yes	No	No	No	No	No	No	No	No	--
Subaru Crosstrek Hybrid 2019	No	Yes	Yes	Yes	Yes	No	Yes	Yes	Yes	No	Yes	Guide recommends not flooding battery and allowing it to burn itself out.
Tesla Model 3 EV 2017–2018	No	Yes	Yes	Yes	Yes	No	Yes	Yes	Yes	No	Yes	Guide states 3,000 gallons of water applied directly to battery may be needed. Consider allowing to burn; can take 24 hours to extinguish.

1 Vehicle make and model	2 ISO 17840 format	3 Elements of NFPA vehicle-specific guides	4 Vehicle-specific instruct-tions for HV disconnect	5 Stabiliza-tion and PPE information	6 General informa-tion on HV battery damage or fire	7 Vehicle-specific instruct-tions for HV battery fire	8 Submer-ged vehicle informa-tion	9 Damaged EV recovery/ towing informa-tion	10 Damaged EV storage informa-tion	11 Informa-tion on mitigating stranded energy risk	12 HV battery characteristics	13 Notes
Toyota Prius PHEV 2012–2015	No	Yes	Yes	Yes	Yes	No	Yes	Yes	Yes	No	Yes	Guide recommends allowing fire to burn itself out.
Van Hool Bus (fuel cell)	No	No	Yes	No	No	No	No	No	No	No	No	Bus has both hydrogen fuel cell and lithium-ion batteries.
Volkswagen e-Golf EV 2014–2018	No	No	Yes	No	No	No	No	No	No	No	No	Only 3-page quick response guide is provided.
Volvo XC60 PHEV 2018	No	No	Yes	No	No	No	No	No	No	No	No	--
TOTAL Yes	0	28	36	24	28	2	26	21	17	0	30	--
TOTAL No	36	8	0	12	8	34	10	15	19	36	6	--

NOTE: EV = electric vehicle; HV = high voltage.

[a] Manufacturer is no longer in business.

[b] Scion was discontinued in 2016.

Vehicle models that use NiMH rather than lithium-ion batteries were excluded from the review (NiMH batteries do not contain a flammable electrolyte). The number of vehicle models produced by different manufacturers varied from just 1 to more than 20. When a manufacturer published response guides for several vehicle models and model years, the most recent model was selected, with priority given to BEVs over PHEVs (the five PHEVs listed in table 3 contain lithium-ion batteries). If a parent company (such as Toyota) published a guide covering multiple manufacturers or brands (such as Lexus and Toyota) as well as a guide for each specific brand, both types of guide were reviewed.

As shown in table 3, most (28) of the manufacturers' guides contain elements of the vehicle-specific guidance that is part of the NFPA's emergency response guide. However, none of the manufacturers follow the format or order of elements in ISO 17840 in their guides, as recommended by SAE J2990.[87] All manufacturers give vehicle-specific information for disabling the high-voltage system—which means disconnecting the high-voltage circuits from the rest of the vehicle, not deenergizing the high-voltage battery itself. Various devices and procedures are described, such as cut loops for severing the connection between the high-voltage system and the rest of the vehicle, or plugs that can be manually lifted out to disconnect the high-voltage system. The guides generally include diagrams or photos of the disconnect procedures, accompanied by written instructions.

Two-thirds (24) of the manufacturers' guides give information about stabilizing a damaged vehicle and also describe the PPE appropriate for various emergency situations. All but 10 guides contain information about submerged vehicles, such as that the chassis will not be energized if the vehicle is under water (and is therefore safe to touch). Twenty-one give information about towing the vehicle—for example, that a flatbed truck should be used.

Twenty-eight of the manufacturers' guides warn about the hazards, including fire, posed by damaged high-voltage batteries. However, only two manufacturers give vehicle-specific instructions for fighting a high-voltage battery fire (Mitsubishi recommends flooding a damaged vehicle with water; Proterra, which manufactures an electric bus, instructs firefighters to focus water at the battery enclosure). Subaru recommends not flooding a battery but letting it burn itself out. Toyota also recommends letting a battery fire burn out by itself. Hyundai and Proterra recommend monitoring a damaged battery

[87] See section 4.4.

using a thermal camera. Generally missing from the guidance is specific information about where to apply water to a burning high-voltage battery and how to determine when it is safe to stop applying water, such as appropriate temperature thresholds.

None of the manufacturers give specific instructions, such as a list of procedures, for minimizing the dangers posed by stranded energy, including the risk of battery reignition. Six do not provide any information about such battery characteristics as peak voltage.

5. Analysis

This safety report begins with investigations by the NTSB of four electric vehicle fires. Three of the fires (Lake Forest, Mountain View, and Fort Lauderdale) resulted from high-speed, single-vehicle collisions that damaged the vehicles' high-voltage lithium-ion batteries. The fourth fire (West Hollywood) was caused by internal failure of the vehicle's high-voltage lithium-ion battery during normal driving operations.

The three crashes were much more severe and resulted in greater damage to vehicles and batteries than the conditions under which electric vehicles are crash-tested to meet federal safety standards (FMVSS 305 and FMVSS 208). Two of the crashes involved vehicles traveling at well above the posted speed limit (82 mph in Lake Forest and 116 mph in Fort Lauderdale). In the Mountain View crash, the vehicle hit a nonoperational crash attenuator at highway speeds (71 mph). All three crashes caused extensive damage that extended into the protected area of the high-voltage battery case, rupturing the case and damaging the battery modules and individual cells.

The Mountain View and Fort Lauderdale crashes resulted in the deaths of vehicle occupants. FMVSS 305 requirements (to prevent electrolyte spillage, retain electrical storage or conversion devices, and ensure electric safety) are evaluated after an electric vehicle is crash-tested under the conditions described in FMVSS 208 (focused on occupant protection at impact speeds ranging from 30 to 50 mph). Both occupants survived the Lake Forest crash, illustrating that crashes at higher impact speeds that cause severe battery damage and fire can be survivable.

As part of its analysis, the NTSB examined national and international standards established to maximize the safety of electric vehicles. Particular attention was paid to the guidance supplied by electric vehicle manufacturers for first and second responders to vehicle crashes and high-voltage battery fires. The examination focused on high-voltage lithium-ion batteries. The analysis identified two primary safety issues:

- Inadequacy of vehicle manufacturers' emergency response guides for minimizing the risks to first and second responders posed by high-voltage lithium-ion battery fires in electric vehicles.
- Gaps in safety standards and research related to high-voltage lithium-ion batteries involved in high-speed, high-severity crashes.

Below, we discuss the findings of the investigation and present recommendations for improving the safety of first and second responders who deal with high-voltage lithium-ion battery fires in electric vehicles.

5.1. Guidance for Emergency Responders

The NTSB reviewed the guidance documents prepared for emergency responders by 36 manufacturers of electric vehicles powered by high-voltage lithium-ion batteries. The results, tabulated in table 3, section 4.5.2, show that the guidance is strong in certain respects but lacking in others, particularly as regards specific instructions for fighting high-voltage lithium-ion battery fires and for addressing the risks posed by stranded energy. Each subsection below draws conclusions related to one critical aspect of the guidance, followed by a subsection containing recommendations for improving the guidance and for disseminating information to first and second responders about the risks and the guidance described in this report.

5.1.1. High-Voltage Disconnect

Emergency responders face a risk of electric shock from the energy stored in damaged high-voltage lithium-ion batteries. To protect against electric shock, safety standards require isolating an electric vehicle's high-voltage battery system from the chassis, giving minimal values for isolation resistance. They also require specific markings for high-voltage systems and orange coverings to identify high-voltage cables and other components.

To lessen the high-voltage risks in damaged electric vehicles, manufacturers include methods of isolating the high-voltage system by disconnecting the high-voltage circuits between the lithium-ion battery and the powertrain. First responders who consult any of the emergency response guides reviewed by the NTSB will find instructions on how to isolate an electric vehicle's high-voltage lithium-ion battery (refer to column 4, table 3). Each of the emergency response guides includes a graphic illustration of the disconnect method for that vehicle, and in all the electric vehicles covered by the guides, the color orange is used to identify high-voltage components. The methods of disconnecting high-voltage circuits vary by vehicle manufacturer— federal standards (FMVSS 305) do not mandate a uniform method. As noted earlier, although the disconnects will isolate a high-voltage lithium battery, they will not remove energy from the battery itself.

SAE J2990, which addresses the hazards faced by first and second responders to crashes and other incidents involving electric vehicles, recommends that manufacturers incorporate at least two methods of disconnecting the vehicles' high-voltage systems. The recommendation lists the following methods: (1) automatic shutdown; (2) switching the ignition switch to OFF (to disconnect the high-voltage system from the high-voltage sources and discharge the system to ≤ 60 volts DC or 30 volts AC within 10 minutes); (3) cut or disconnect battery cables to discharge the 12-volt system, and cut or disconnect the 12-volt output cable. SAE J2990 states that using a manual disconnect (such as pulling a plug) should not be the primary method for first responders to disable a vehicle's high-voltage circuits. That is because (1) the variety of designs makes locating and activating manual disconnects inefficient, (2) first responders do not always have the required PPE, and (3) the manual disconnect mechanism might be inaccessible.

The disabling methods incorporated by manufacturers include manual service disconnects and cut loops that can be severed to disconnect the high-voltage system and isolate it from the rest of the vehicle. A few manufacturers prefer operating the manual disconnect as the method of disabling the high-voltage system. If physical disconnects are not available or are inaccessible, some manufacturers recommend pulling a particular fuse, or removing all fuses. For several models, the guidance states that the vehicle's high-voltage system will automatically disconnect if the airbags deploy. For other models, emergency responders are told that pressing the start/stop button will automatically trigger a high-voltage disconnect.

All the vehicles in the three postcrash battery fires investigated for this report were equipped with emergency disconnect devices. Under the circumstances, however, none of the first responders to the fires employed any of the devices. In the Lake Forest case, the high-voltage cut loop was in the area of severe fire and impact damage at the front of the SUV. The Mountain View crash resulted in the cut loop (also at the front of the vehicle) being completely torn away. During the inspection after the Fort Lauderdale crash, firefighters told NTSB investigators that the fire made the cut loop at the front of the car inaccessible, even if they had known where to look for it.[88] In the West Hollywood fire, which did not follow a crash, first responders successfully cut the vehicle's high-voltage disconnect loop, though they expressed confusion about the relation between the high-voltage system and the 12-volt system that powered the vehicle's computer. The NTSB concludes that manufacturers' emergency response guides provide sufficient vehicle-specific information for disconnecting an electric vehicle's high-voltage system when the high-voltage disconnects are accessible and undamaged by crash forces. The NTSB further concludes that crash damage and resulting fires may prevent first responders from accessing the high-voltage disconnects in electric vehicles.

5.1.2. Fire Suppression

The NFPA emergency field guide states that large, sustained volumes of water are required to extinguish a high-voltage battery fire: "it could require over 2,600 gallons, depending on the size and location of the battery." The guidance also highlights the difficulty of applying extinguishing agents directly onto burning cells because of the batteries' protective cases. It further states that applying a large volume of water might cool the battery enough to prevent the fire from propagating to adjacent cells. A high-voltage lithium-ion battery is designed to resist water, but water is critical for cooling overheated cells to stop thermal runaway and further combustion (as discussed in the next section).

In the four NTSB investigations, the total amount of water used to suppress the high-voltage lithium-ion battery fires ranged from 300 gallons in Fort Lauderdale and West Hollywood to 1,000 gallons in Mountain View and to over 20,000 gallons in Lake Forest. In the Fort Lauderdale incident, the battery fire was intense, but a rupture in the battery case between the front seats allowed water into the case, which helped extinguish the fire. In the West Hollywood battery fire, which did not result from a crash, the flames were extinguished quickly, but firefighters had to apply water for a further 30 minutes to stop the battery from smoking. In the Mountain View case, the flames were also extinguished quickly, but more water had to be applied when the battery fire reignited. In the Lake Forest case, firefighters applied large

[88] Cut loops can be in the front or the rear, or in both front and rear, as in Tesla model S cars built after June 2013 (such as the vehicle in the Fort Lauderdale fire).

quantities of water, but they could not extinguish the fire until they elevated the vehicle and applied water directly to the battery on the underside.

According to the NTSB's review of the emergency response guides available to first and second responders, most manufacturers provide general information about damaged high-voltage batteries and the danger of fire, including that such fires require large amounts of water to extinguish. However, manufacturers generally do not give vehicle-specific information for fighting a high-voltage lithium-ion battery fire (refer to column 7, table 3), such as where and how to apply water in order to extinguish and cool the battery. In the Lake Forest fire, firefighters searched online for assistance, but they did not apply enough water directly onto the battery case in the beginning to cool the high-voltage battery, which ended up requiring 2 hours and 20,000 gallons of water to extinguish. In the West Hollywood fire, firefighters contacted the manufacturer directly once the flames had died down because they could not determine where to apply water to stop the vehicle from smoking. While searching for a way to apply water to the battery, firefighters used a metal pry bar to remove body panels from the vehicle, a procedure that posed a risk of electric shock and that guidance documents warn against.

Thus, in addition to hampering efficient extinguishing of high-voltage lithium-ion battery fires, the lack of clear, vehicle-specific firefighting information can lead to confusion or inadvisable action on the part of first responders, even when general guidance is available. The NTSB concludes that the instructions in most manufacturers' emergency response guides for fighting high-voltage lithium-ion battery fires lack necessary, vehicle-specific details on suppressing the fires.

The NTSB notes that in France, Renault (not included in the guidance documents reviewed for this report because its vehicles are not imported to the United States) has worked with fire and emergency services to design inlet ports through which, in case of fire, water can be applied directly to the lithium-ion batteries in its electric vehicles. According to a company website, firefighters can extinguish a battery fire in less than a minute using the access ports.[89]

5.1.3. Thermal Runaway and Battery Reignition

Damaged high-voltage lithium-ion batteries pose a risk to emergency responders because of the potential for thermal runaway, which can cause a

[89] See article dated May 16, 2018, on the Groupe Renault website: "Electric Vehicles: Groupe Renault Works Hand in Hand with Fire Services" (accessed June 1, 2020).

battery to ignite, or reignite. The risk of thermal runaway in electric vehicles powered by high-voltage lithium-ion batteries became evident in 2011, when a Chevrolet Volt caught fire 3 weeks after a crash test.

The high-voltage energy stored in lithium-ion batteries and the flammability of the batteries' electrolyte create the potential for thermal runaway and fire. Thermal runaway begins with overheating in individual battery cells that degrades their electrical isolation and can be in progress even when flames are not visible outside the battery. SAE J2990 recommends using a thermal camera or infrared temperature probe to inspect a damaged battery. Emergency responders are also counseled to use their senses when inspecting a damaged electric vehicle—to listen for noises from the battery (gurgling, bubbling, crackling, hissing, or popping), which can indicate that overheated cells are venting or that the high-voltage system is arcing; and to notice whether a burnt odor is coming from the battery, which is evidence of fire or heat damage.

SAE J2990 warns that if a high-voltage lithium-ion battery is involved in a fire, "there is a possibility that it could reignite after extinguishment." The NFPA's emergency field guide also warns that a fire in a high-voltage lithium-ion battery could reignite and recommends that emergency responders use thermal imaging to monitor the battery. The guide states that reignition is accompanied by a "whooshing" or "popping" sound, followed by off-gassing of white smoke or electrical arcs or sparks that reignite, with visible flames or fire. The guidance states that reignition can occur within several hours to a day or more after the fire is extinguished. The NFPA advises firefighters to continue applying water (even after they can no longer see a flame) to sufficiently cool the battery pack—it could take an hour or more—so as to reduce the risk of reignition.

All but 8 of the 36 manufacturers' emergency response guides reviewed for this report contain general information about high-voltage battery damage, including that high-voltage batteries can reignite after being extinguished. Tesla's guidance for the model S specifically states that about 3,000 gallons of water are needed to cool the high-voltage lithium-ion battery, that reignition can occur, and that the battery must be completely cooled before it is released to secondary responders. The vehicles in both the Lake Forest and Mountain View crashes experienced at least six battery reignitions after the vehicle fires were extinguished. The battery in the Fort Lauderdale crash reignited three times. (The battery in the West Hollywood fire did not reignite.) Emergency responders in all three cases either consulted the emergency response guides online or contacted the manufacturer directly. Yet firefighters and tow truck drivers were unable to prevent the reignitions. The NTSB concludes that thermal runaway and multiple battery reignitions after initial fire suppression are safety risks in high-voltage lithium-ion battery fires.

5.1.4. Stranded Energy

The NTSB concludes in section 5.1.1 that the manufacturers' emergency response guides contain sufficient information for first responders to disconnect an electric vehicle's high-voltage system, provided the disconnects are accessible and are undamaged by crash forces. However, even if an electric vehicle's high-voltage system is disconnected, energy will remain trapped in a damaged high-voltage lithium-ion battery. The stranded energy poses a risk of electric shock to emergency responders and creates the potential for thermal runaway that can result in reignition and fire.

Battery reignition can be caused by thermal or mechanical means. In a thermal reignition, the battery is not sufficiently cooled and thermal runaway results, as in the Lake Forest fire. Mechanical reignition results when parts of a battery or other conductive debris cause a short circuit in cells that contain stranded energy. For example, if a damaged electric vehicle is shifted, twisting of the wreckage can create new electrical connections that can then release energy and cause reignition. After the Fort Lauderdale fire, the battery reignited twice—once while the wreckage was being loaded onto a tow truck, and again when a chain passed over the battery after the vehicle had been loaded onto the truck.

In the three high-speed, high-severity crashes that the NTSB investigated, the vehicles' high-voltage lithium-ion batteries reignited at least 15 times in total. The reignitions were evidence that stranded energy remained in undamaged parts of the high-voltage batteries. In the inspection of the battery system after the Fort Lauderdale crash and fire, the voltage in battery modules 1 through 5 ranged from 69.9 to 167.3 volts DC, well above the safe limits for human exposure (50 to 60 volts DC). The inspection of the high-voltage battery after the West Hollywood fire found that module 14 had experienced internal failure and that all other modules remained charged.

Technicians at a local service center operated by the manufacturer, which took possession of the vehicle, removed the battery and discharged the stranded energy by 50 percent.[90]

---

[90] As noted earlier, the charge was reduced to 50 percent to allow safe long-distance transportation of the battery and also to preserve evidence.

After the Mountain View crash, engineers twice attempted to remove stranded energy from the vehicle's high-voltage energy system. On scene, they halted their attempt when they encountered signs of reignition. After the vehicle had been moved to the more-controlled environment of a tow yard, the engineers made a second attempt to remove stranded energy from the battery, using a connector on the battery's outside. However, the connector was contaminated by water and debris, and attempts to deenergize the battery failed. Efforts to measure the voltage were inconclusive.

After the Lake Forest crash and fire, much of the high-voltage lithium-ion battery remained intact, but confirming that the battery contained stranded energy was not possible because the module connection terminals were inaccessible. In the Chevrolet Volt example from 2011, sufficient stranded energy remained after a crash test that the battery caught fire while the vehicle was in storage 3 weeks later. In two international examples of high-voltage lithium-ion battery fires in electric vehicles reviewed for this study, a BMW model i3 and a Mitsubishi Outlander also experienced reignitions. Although none of the battery fires investigated or reviewed for this safety report resulted in electric shock to vehicle occupants or emergency responders, an energized high-voltage lithium-ion battery always presents such a risk. The NTSB concludes that the energy remaining in a damaged high-voltage lithium-ion battery, known as stranded energy, poses a risk of electric shock and creates the potential for thermal runaway that can result in battery reignition and fire.

Firefighters have no method of determining whether stranded energy is present in a damaged high-voltage lithium-ion battery or of removing energy from the battery pack.[91] Engineers or other specialists can use the battery management system to check for remaining voltage if the system is operational, and some batteries have built-in discharge ports. However, in three of the four NTSB investigations discussed in this report, the high-voltage battery system was damaged in a crash, preventing access to the battery management system or to the discharge ports. In addition, the status of a damaged battery is unknown and must be treated as a high-voltage safety risk.

In all four NTSB investigations, emergency response guides for each vehicle were available to the first responders from both the NFPA website and directly from the vehicle manufacturer. The NTSB's review of emergency response guides from 36 manufacturers found that all contained vehicle-

---

[91] Existing methods of deenergizing a battery (such as accessing the terminals manually or flooding the battery pack with a conductive solution) are not within an emergency responder's scope and typically take an impractically long time.

specific information for mitigating high-voltage risks, such as the location of the high-voltage battery and methods for disconnecting the high-voltage system. However, none of the guides offered information for mitigating the risks of stranded energy, such as a list of procedures for minimizing the chance of battery reignition or specific instructions for where and how to apply water to cool a high-voltage battery. One method of removing the stranded energy from a high-voltage lithium-ion battery is to separate it from the vehicle and discharge it in a saltwater bath.

The NFPA's recent training video addressing high-voltage battery systems and stranded energy highlights the challenges to emergency responders.[92] The NFPA warns responders that they should always assume that the high-voltage system is energized. First responders therefore must assume that stranded energy is present in a damaged electric vehicle, and they require vehicle-specific information on how to manage—and minimize—the associated risks for battery reignition and fire.[93] Secondary responders, including tow operators and storage facility operators, must also assess and mitigate the risks associated with high-voltage lithium-ion batteries. As noted earlier, in the Fort Lauderdale recovery effort, a chain that passed over the high-voltage battery caused it to reignite. After the Lake Forest crash, when the vehicle ignited while being moved onto a flatbed tow truck, the tow truck operator suffered minor injuries. The NTSB concludes that high- voltage lithium-ion batteries in electric vehicles, when damaged by crash forces or internal battery failure, present special challenges to first and second responders because of insufficient information from manufacturers on procedures for mitigating the risks of stranded energy.

Storing an electric vehicle with a high-voltage lithium-ion battery is problematic when the vehicle is damaged or a battery fire has occurred. The battery fires investigated by the NTSB reignited multiple times, even days after a crash, placing other vehicles and nearby structures at risk for fire. The Chevrolet Volt that reignited 3 weeks after a crash test damaged four nearby vehicles. The NFPA and vehicle manufacturers recommend leaving a 50-foot clearance around a stored, damaged electric vehicle to prevent fire damage to nearby vehicles or structures in case the battery reignites. (The guidance in SAE J2990 gives an alternate solution: create a barrier of earth, steel, concrete, or solid masonry around the vehicle.) The storage facilities for the vehicles

---

[92] The video can be viewed on the NFPA's training and certification website (accessed June 5, 2020).

[93] The NFPA video highlights the need for responders to consider the type, quantity, and location of batteries in a damaged electric vehicle.

involved in the NTSB-investigated crashes and fires in Lake Forest, Mountain View, and Fort Lauderdale could not accommodate a 50-foot radius around the damaged vehicles. If more than one electric vehicle with a risk for fire reignition were stored at the same facility, providing sufficient space to mitigate the risks of a subsequent fire would most likely be impossible. The NTSB concludes that storing an electric vehicle with a damaged high-voltage lithium-ion battery inside the recommended 50-foot-radius clear area may be infeasible at tow or storage yards.

5.1.5. Format Issues

Emergency response guides must be clear about the vehicle-specific procedures and evaluation criteria necessary to minimize the risk of ignition or reignition of a damaged high-voltage lithium-ion battery. The fire/rescue organization CTIF developed ISO standard 17840 (*Road Vehicles—Information for First and Second Responders*) as a four-part standard addressing all vehicle types, including passenger cars, light commercial vehicles, buses, and heavy commercial vehicles (see section 4.4 for details). Part 3 of the standard, published in April 2019, contains a template for emergency response guides that is applicable to various types of vehicles, including electric vehicles. The template defines the layout and general contents of emergency response guides, with the goal of establishing a standard format for clarity and ease of use by first and second responders. The aim is that responders will have the information they need to make quick, correct decisions in case of fire, vehicle submersion, or fluid leakage. The standard spells out the headings for the various sections of an emergency response guide (such as "Disable direct hazards/safety regulations" or "In case of fire"), but leaves it to manufacturers to fill in specific information for each section of the guide.

The NTSB found that none of the electric vehicle manufacturers organize their emergency response guides according to the format laid out in ISO standard 17840, as recommended in SAE J2990. The ISO standard is intended to support the quick and safe rescue of vehicle occupants as well as the safety of emergency responders. The benefits of standardized guidance documents also extend to enhancing the training of emergency responders. If manufacturers used the ISO standard 17840 template to present emergency response information, it could increase response efficiency, improve emergency responders' understanding of actions to take when confronting a damaged electric vehicle and its high-voltage lithium-ion battery, enable consistent training of emergency responders, and minimize the risks to emergency responders associated with high- voltage energy systems, because the information would be presented in a consistent manner for all vehicles and contain equivalent levels of vehicle-specific detail. The NTSB therefore concludes that electric vehicle manufacturers should use the ISO standard 17840 template to present emergency response information. Euro NCAP, in coordination with CTIF, is incorporating scoring relative to the availability of a manufacturer's emergency response guide and its compliance with ISO standard 17840.[94] The NTSB concludes that action by NHTSA, similar to that taken by Euro NCAP, to incorporate scoring relative to the availability of a manufacturer's emergency response guide and its adherence to ISO standard 17840 and SAE standard J2990 into the US NCAP, would be an incentive for manufacturers of vehicles sold in the United States with high-voltage battery systems to comply with those standards.

5.1.6. Recommendations for Improving Guidance and Disseminating Information

The NTSB found that the guidance provided by electric vehicle manufacturers to first and second responders is lacking in important respects (as shown by the "no" entries in table 3, section 4.5.2)—particularly as regards the safety risks of fighting fires that involve high-voltage lithium-ion batteries and the dangers posed by the stranded energy in a damaged high-voltage lithium-ion battery. The NTSB identified that nearly all the manufacturers' emergency response guides contain general information about the risks posed by the high-voltage lithium-ion batteries in electric vehicles—such as that damaged batteries can ignite or reignite—but that only two guides offer vehicle-specific information for suppressing a high-voltage battery fire. Critical information includes where to apply water to extinguish a battery fire in an electric vehicle and when it is safe to stop applying water, such as when temperature thresholds are reached.

Of further concern is that none of the reviewed guides give emergency responders any information about how to handle the risks posed by the stranded energy in a damaged high-voltage lithium-ion battery. Firefighters who participate in the initial emergency response and tow truck drivers who move a damaged electric vehicle from the scene can all be exposed to the risks of stranded energy. Storing an electric vehicle with a damaged high-voltage lithium-ion battery is an additional problem. The guidance documents

[94] The scoring is based on the availability of the short versions of manufacturers' emergency response guides, known as rescue sheets.

generally limit their advice to calling the manufacturer for instructions or leaving at least 50 feet of clearance around the vehicle. The NTSB considers that advice impractical at best. Given the SAE's recommendation to also consider isolating a damaged electric vehicle inside a barrier of earth, steel, concrete, or masonry, manufacturers might consider including such an alternative in their guidance.

One obvious way of making the guidance information readily accessible to emergency responders would be to present it in the form of a checklist, similar to the list in table 1, section 4.3.3 (the postincident steps recommended in SAE J2990). The NTSB agrees with SAE and others that having the emergency response guides published in a clear, consistent format would improve their usefulness to emergency responders as well as minimize the risks to responders associated with high-voltage lithium-ion batteries. The NTSB also urges vehicle manufacturers to include in their emergency response guides information specific to each vehicle so that firefighters and tow truck drivers can efficiently and safely act in emergencies involving electric vehicles that are powered by high-voltage lithium-ion batteries.

The NTSB therefore recommends that the manufacturers of electric vehicles equipped with high-voltage lithium-ion batteries model their emergency response guides on ISO standard 17840, as included in SAE recommended practice J2990, and incorporate vehicle-specific information on (1) fighting high-voltage lithium-ion battery fires; (2) mitigating thermal runaway and the risk of high-voltage lithium-ion battery reignition; (3) mitigating the risks associated with stranded energy in high-voltage lithium-ion batteries, both during the initial emergency response and before moving a damaged electric vehicle from the scene; and (4) safely storing an electric vehicle that has a damaged high-voltage lithium-ion battery.[95]

The electric vehicle market is expanding, and NHTSA's NCAP rates the performance of new vehicle models against various safety measures. The NTSB concluded in section 5.1.5 that incorporating scoring into the NCAP

---

[95] The manufacturers (including subdivisions or subsidiary brands) subject to this recommendation are as follows: BMW Group (BMW and MINI); BYD Motors; FCA Group (Chrysler and Fiat); General Motors Company (Buick, Cadillac, Chevrolet, and GMC); Ford Motor Company (Ford and Lincoln); Gillig; Honda Motor Company (Honda and Acura); Hyundai Motor Company; Karma Automotive; Kia Motors Corporation; Mercedes-Benz USA; Mitsubishi Motors; Nissan Motor Company (Infiniti and Nissan); Nova Bus, Inc.; Porsche Cars North America; Proterra, Inc.; North American Subaru; Tesla, Inc.; Toyota Motor North America (Lexus and Toyota); Van Hool NV; Volkswagen Group of America (Audi and Volkswagen); and Volvo Car Corporation. Three companies whose emergency response guides were reviewed for this report have gone out of business: Azure Dynamics, Fisker, and Smith. Toyota discontinued its Scion brand in 2016.

relative to the availability of a manufacturer's emergency response guide and its adherence to the relevant ISO and SAE standards would be an incentive for manufacturers of vehicles sold in the United States with high-voltage battery systems to comply with those standards. The incentive would apply to both current and new manufacturers of electric vehicles, and adherence would enhance the safety of first and second responders by improving the emergency response guides provided by manufacturers. The NTSB therefore recommends that when determining a vehicle's US NCAP score, NHTSA factor in the availability of a manufacturer's emergency response guide and its adherence to ISO standard 17840 and SAE recommended practice J2990.

It is vital that first and second responders be aware of the risks associated with the high-voltage lithium-ion batteries in electric vehicles. The NFPA and other organizations have been working to disseminate such information. The NTSB is particularly encouraged that the dangers of stranded energy have begun to be publicized, as in the NFPA's new training video and recent journal article (Roman 2020). Nevertheless, some firefighters and tow truck drivers may not fully understand the risks entailed in fighting a high-voltage battery fire or in transporting a damaged electric vehicle, or they may not know where to find guidance information. An example of responders' concerns that reflect a lack of clear understanding occurred during efforts to suppress the final battery reignition after the Mountain View crash. Firefighters applied water sporadically, fearing electric shock from the water stream—a potential risk the NFPA has studied and discounted.

The professional associations that represent or operate training programs for first and second responders include, in addition to the NFPA, the International Association of Fire Chiefs, representing the leadership of firefighters and emergency responders worldwide; the International Association of Fire Fighters, a labor union representing full-time firefighters and emergency medical personnel in the United States and Canada; the National Alternative Fuels Training Consortium, a nationwide training organization that develops curriculums and implements training, outreach, and education on the use of alternative fuel vehicles; the National Volunteer Fire Council, which represents volunteer fire, emergency medical, and rescue services and administers the National Traffic Incident Management Responder Training Program; and the Towing and Recovery Association of America, which, in partnership with the Federal Highway Administration, operates the National Driver Certification Program for tow truck operators. The NTSB therefore recommends that the NFPA, the International Association of Fire Chiefs, the International Association of Fire Fighters, the National Alternative

Fuels Training Consortium, the National Volunteer Fire Council, and the Towing and Recovery Association of America inform their members about the circumstances of the fire risks described in this report and the guidance available to emergency personnel who respond to high-voltage lithium-ion battery fires in electric vehicles.

5.2. Standards and Research

The need to evaluate the risks associated with high-voltage lithium-ion battery fires, stranded energy, and postcrash (or postincident) vehicle storage is beginning to be recognized. In 2017, NHTSA published a report addressing safety issues in the high-voltage lithium-ion battery systems of electric vehicles and identified crash and postcrash damage as a gap in regulations and safety standards (Stephens and others 2017). Since that report, NHTSA has initiated research projects into battery management systems and stranded energy. For example, Rask and others (2020) reviewed the risks and hazards associated with the stranded energy remaining in high-voltage lithium-ion battery systems. The research focused on undamaged vehicle battery failures or vehicles damaged in survivable crashes (FMVSS 208 level). The mitigation strategies (which included manufacturer-developed battery discharge tools) assumed access to battery connections, which, as seen in the NTSB-investigated crashes, might not be feasible on scene or not possible at all after a high-speed, high-severity crash. Further, the current federal standard (FMVSS 305), the efforts of the United Nations working group on electric vehicle safety (GTR 20), and the emergency response guidance detailed in SAE J2990 do not currently address high-speed, high-severity crashes resulting in damage to a high-voltage lithium-ion battery and the associated stranded energy in the high-voltage system. The NTSB concludes that although existing standards address damage sustained by high-voltage lithium-ion battery systems in survivable crashes, as defined by federal crash standards, they do not address high-speed, high-severity crashes resulting in damage to high-voltage lithium-ion batteries and the associated stranded energy.

NHTSA is best positioned to facilitate research in the United States on high-speed, high-severity vehicle crashes that result in damage to high-voltage lithium-ion batteries and to communicate findings to the international electric vehicle community. One goal of the Rask and others (2020) study was to develop a prototype tool that would enable nonexperts, such as tow truck drivers, to assess and possibly deenergize a high-voltage battery system after a crash. However, the prototype tool would require either a functional battery management system or a direct connection to internal battery modules through a high-voltage port. The battery management system or the ports, or both, are likely to be damaged in a crash or in a thermal runaway event (as seen in the NTSB investigations discussed in this report), making access to them problematic or impossible. Existing deenergizing devices take hours or days to remove the stranded energy from an undamaged high-voltage lithium-ion battery. The NTSB therefore recommends that NHTSA convene a coalition of stakeholders to continue research initiated by NHTSA on ways to mitigate or deenergize the stranded energy in high-voltage lithium-ion batteries and to reduce the hazards associated with thermal runaway resulting from high-speed, high-severity crashes. NHTSA should publish the research results.

6. Findings

1. Manufacturers' emergency response guides provide sufficient vehicle-specific information for disconnecting an electric vehicle's high-voltage system when the high-voltage disconnects are accessible and undamaged by crash forces.
2. Crash damage and resulting fires may prevent first responders from accessing the high-voltage disconnects in electric vehicles.
3. The instructions in most manufacturers' emergency response guides for fighting high-voltage lithium-ion battery fires lack necessary, vehicle-specific details on suppressing the fires.
4. Thermal runaway and multiple battery reignitions after initial fire suppression are safety risks in high-voltage lithium-ion battery fires.
5. The energy remaining in a damaged high-voltage lithium-ion battery, known as stranded energy, poses a risk of electric shock and creates the potential for thermal runaway that can result in battery reignition and fire.
6. High-voltage lithium-ion batteries in electric vehicles, when damaged by crash forces or internal battery failure, present special challenges to first and second responders because of insufficient information from manufacturers on procedures for mitigating the risks of stranded energy.

7. Storing an electric vehicle with a damaged high-voltage lithium-ion battery inside the recommended 50-foot-radius clear area may be infeasible at tow or storage yards.
8. Electric vehicle manufacturers should use the International Organization for Standardization standard 17840 template to present emergency response information.
9. Action by the National Highway Traffic Safety Administration, similar to that taken by the European New Car Assessment Programme, to incorporate scoring relative to the availability of a manufacturer's emergency response guide and its adherence to International Organization for Standardization standard 17840 and SAE International recommended practice J2990 into the US New Car Assessment Program, would be an incentive for manufacturers of vehicles sold in the United States with high-voltage lithium- ion battery systems to comply with those standards.
10. Although existing standards address damage sustained by high-voltage lithium-ion battery systems in survivable crashes, as defined by federal crash standards, they do not address high-speed, high-severity crashes resulting in damage to high-voltage lithium-ion batteries and the associated stranded energy.

7. Recommendations

As a result of its investigation, the National Transportation Safety Board makes the following new safety recommendations:

To the National Highway Traffic Safety Administration:

When determining a vehicle's US New Car Assessment Program score, factor in the availability of a manufacturer's emergency response guide and its adherence to International Organization for Standardization standard 17840 and SAE International recommended practice J2990. (H-20-30)

Convene a coalition of stakeholders to continue research initiated by your organization on ways to mitigate or deenergize the stranded energy in high-voltage lithium-ion batteries and to reduce the hazards associated with thermal runaway resulting from high-speed, high-severity crashes. Publish the research results. (H-20-31)

To the manufacturers of electric vehicles equipped with high-voltage lithium-ion batteries: (BMW Group; BYD Motors; FCA Group; General Motors Company; Ford Motor Company; Gillig; Honda Motor Company; Hyundai Motor Company; Karma Automotive; Kia Motors Corporation; Mercedes-Benz USA; Mitsubishi Motors; Nissan Motor Company; Nova Bus, Inc.; Porsche Cars North America; Proterra, Inc.; North American Subaru; Tesla, Inc., Toyota Motor North America; Van Hool NV; Volkswagen Group of America; and Volvo Car Corporation)

Model your emergency response guides on International Organization for Standardization standard 17840, as included in SAE International recommended practice J2990, and incorporate vehicle-specific information on (1) fighting high-voltage lithium-ion battery fires; (2) mitigating thermal runaway and the risk of high-voltage lithium-ion battery reignition; (3) mitigating the risks associated with stranded energy in high-voltage lithium-ion batteries, both during the initial emergency response and before moving a damaged electric vehicle from the scene; and (4) safely storing an electric vehicle that has a damaged high-voltage lithium-ion battery. (H-20-32)

To the National Fire Protection Association, the International Association of Fire Chiefs, the International Association of Fire Fighters, the National Alternative Fuels Training Consortium, the National Volunteer Fire Council, and the Towing and Recovery Association of America

Inform your members about the circumstances of the fire risks described in this report and the guidance available to emergency personnel who respond to high-voltage lithium-ion battery fires in electric vehicles. (H-20-33)

By The National Transportation Safety Board

ROBERT L. SUMWALT, III
Chairman

JENNIFER HOMENDY
Member

BRUCE LANDSBERG
Vice Chairman

MICHAEL GRAHAM
Member

THOMAS B. CHAPMAN
Member

Report Date: November 13, 2020

Appendix: Lithium-Ion Battery Fires in Aircraft

Lithium-ion batteries can power aircraft systems as well as automobiles. In 2013, the NTSB conducted three investigations of lithium-ion batteries as installed aircraft equipment and hosted a forum on lithium-ion batteries in transportation. In 2020, the NTSB issued a safety recommendation report addressing the transportation of lithium-ion batteries by air.

Investigations

The NTSB investigated a January 7, 2013, battery fire in the auxiliary power unit of a Japan Airlines Boeing 787-8 in Boston.[96] The plane was parked at a Boston airport when maintenance personnel saw smoke and fire coming from the battery case of the aircraft's auxiliary power unit. The NTSB concluded that the probable cause of the fire was an internal short circuit in the power unit's lithium-ion battery, which led to thermal runaway that cascaded to adjacent cells and the release of smoke and fire. NTSB safety recommendations addressed assessing and managing the risk of short circuit and fire in lithium-ion batteries.

The NTSB assisted the Japan Transport Safety Board with its investigation of a battery incident on an All Nippon Airways Boeing 787 on

[96] For further information, see NTSB (2014) Also, https://www.ntsb.gov/investigations/Pages/boeing_787.aspx.

January 16, 2013.[97] The Japan Transport Safety Board concluded that a cell in the aircraft's main battery had most likely suffered a short circuit, resulting in thermal runaway that destroyed the battery and led to emergency landing and evacuation of the aircraft. The NTSB also assisted the General Civil Aviation Authority of the United Arab Emirates with its investigation of the cargo fire and crash of a United Parcel Service Boeing 747-400F on September 3, 2013, in the desert outside Dubai. The General Civil Aviation Authority's investigation concluded "with reasonable certainty that the location of the fire was in an element of the cargo that contained, among other items, lithium batteries."[98]

Forum

In April 2013, the NTSB convened a public forum titled "Lithium Ion Batteries in Transportation."[99] Panels discussed the design, development, and performance of lithium-ion batteries; regulations and standards associated with the manufacture of lithium-ion batteries, consumer and industry use, and transportation of the batteries as cargo; and the application and safety aspects of lithium-ion battery technology in various transportation modes.

Safety Recommendation Report

In May 2020, the NTSB issued a safety recommendation report titled Standards for Lithium-Ion Battery Shipments by Air (NTSB 2020b). The report followed an accident on June 3, 2016, in which a FedEx delivery truck and all its cargo were destroyed by fire while the driver was delivering four large-format lithium-ion batteries to a business in Brampton, Ontario, Canada. No injuries were reported. The batteries had been transported from Sarasota, Florida, in two US-registered cargo planes. The probable cause of the fire was

---

[97] For further information on the Nippon Airways fire, see the Japan Transport Safety Board's report: Emergency Evacuation Using Slides, All Nippon Airways Co., Ltd., Boeing 787-8, JA804A, Takamatsu Airport, At 08:49 JST, January 16, 2013. Aircraft Serious Incident Investigation Report AI2014-4 (Tokyo, Japan: JTSB) (accessed November 12, 2020).

[98] For further information, see the United Arab Emirates report: Uncontained Cargo Fire Leading to Loss of Control Inflight and Uncontrolled Descent Into Terrain, Final Air Accident Investigation Report, Boeing 747-44AF, N571UP, case 13/2010 (Dubai, United Arab Emirates: Air Accident Investigation Sector) (accessed November 12, 2020).

[99] See the NTSB event summary.

determined to be an electrical short circuit in one of the batteries, causing a thermal runaway that ignited the battery and its packaging. The NTSB concluded that if the thermal runaway had occurred on an airplane, the accident could have resulted in significant damage to or loss of the airplane. The NTSB made recommendations to the Pipeline and Hazardous Materials Safety Administration to remove special provisions and exemptions from testing for low-production or prototype lithium-ion batteries that are transported by air.

References

Bisschop. R., O. Willstrand, F. Amon, and M. Rosengren. 2019. *Fire Safety of Lithium-Ion Batteries in Road Vehicles.* Report 2019:50. Gothenburg: RISE Research Institutes of Sweden.

Grant, C. C. 2010. *Fire Fighter Safety and Emergency Response for Electric Drive and Hybrid Electric Vehicles.* Final report. Quincy, Massachusetts: Fire Protection Research Foundation.

Kane, M. 2018. "Hotstick Reveals High Voltage Detector for First Responders." *InsideEVs*.June 29.

Long, R. T., Jr., A. F. Blum, T. J. Bress, and B.R.T. Cotts. 2013. *Best Practices for Emergency Response to Incidents Involving Electric Vehicle Battery Hazards: A Report on Full-Scale Testing Results*. Final report (part 1, part 2, part 3). Prepared for Fire Protection Research Foundation, Quincy, Massachusetts, by Exponent, Inc., Bowie, Maryland.

NFPA (National Fire Protection Association). 2018. *NFPA's Alternative Fuel Vehicles Safety Training Program: Emergency Field Guide, 2018 Edition.* Quincy, Massachusetts: NFPA.

NTSB (National Transportation Safety Board). 2014. *Auxiliary Power Unit Battery Fire, Japan Airlines Boeing 787-8, JA829J, Boston, Massachusetts, January 7, 2013.* Incident Report NTSB/AIR-14/01. Washington, DC: NTSB.

. 2019a. *Addressing Systemic Problems Related to the Timely Repair of Traffic Safety Hardware in California.* Highway Safety Recommendation Report NTSB/HSR-19/01. Washington, DC: NTSB.

. 2019b. *Single-Vehicle Run-Off-Road Crash and Fire, Fort Lauderdale, Florida, May 8, 2018.* Highway Accident Brief NTSB/HAB-19/08. Washington, DC: NTSB.

. 2019c. *Air Products and Chemicals, Inc. Tube Trailer Module Hydrogen Release and Subsequent Fire, Diamond Bar, California, February 11, 2018.* Hazardous Materials Incident Report NTSB/HZM-19/02. Washington, DC: NTSB.

. 2020a. *Collision Between a Sport Utility Vehicle Operating With Partial Driving Automation and a Crash Attenuator, Mountain View, California, March 23, 2018.* Highway Accident Report NTSB/HAR-20/01. Washington, DC: NTSB.

. 2020b. *Standards for Lithium-Ion Battery Shipments by Air*. Safety Recommendation Report NTSB/HZMSR-20/01. Washington, DC: NTSB.

Rask, E., C. Pavlich, K. Stutenberg, M. Duoba, and G. Keller. 2020 *Stranded Energy Assessment Techniques and Tools*. DOT HS 812 789. Washington, DC: prepared for NHTSA by Plaza Argonne National Laboratory.

Roman, J. 2020. "Stranded Energy." *NFPA Journal*. January–February.

Smith, B. 2012. *Chevrolet Volt Battery Incident Overview Report*. DOT HS 811 573. Washington, DC: NHTSA.

Stephens, D., P. Shawcross, G. Stout, E. Sullivan, J. Saunders, S. Risser, and J. Sayre. 2017. *Lithium- ion Battery Safety Issues for Electric and Plug-in Hybrid Vehicles*. DOT HS 812 418. Washington, DC: prepared for NHTSA by Battelle Memorial Institute.

UNECE (United Nations Economic Commission for Europe). 2013. Addendum 13: *Global Technical Regulation on Hydrogen and Fuel Cell Vehicles*. ECE/Trans/180/Add. 13. Geneva, Switzerland: United Nations.

. 2018. Addendum 20: Global Technical Regulation No. 20: *Global Technical Regulation on Electric Vehicle Safety (EVS)*. ECE/Trans/180/Add. 20. Geneva, Switzerland: United Nations.

Chapter 2

Examining Fire Hazards: Lithium-Ion Batteries and Other Threats to Fire Safety*

Committee on Homeland Security

Statement of Lori Moore-Merrell, DrPH, MPH
U.S. Fire Administrator
Federal Emergency Management Agency
U.S. Department of Homeland Security,
Before the Committee on Homeland Security Subcommittee
On Emergency Management and Technology
U.S. House of Representatives Washington, D.C.
"Examining Fire Hazards: Lithium-Ion Batteries and Other Threats to Fire Safety" Submitted by Federal Emergency Management Agency

Chairman D'Esposito, Ranking Member Carter, and Members of the Subcommittee:

My name is Lori Moore-Merrell and I serve as the Administrator of the United States Fire Administration (USFA) within the Federal Emergency Management Agency (FEMA). Thank you for the opportunity to testify today and to discuss the continuous and evolving fire threats to the nation.

* This is an edited, reformatted and augmented version of Hearing of Committee on Homeland Security Publication, dated Thursday, February 15, 2024.

In: Lithium-Ion Batteries
Editor: Gregory R. Simmons
ISBN: 979-8-89530-773-1

The USFA's mission is to support and strengthen fire and emergency medical services to prevent, mitigate, prepare for, and respond to all hazards. Since 1974, the USFA has led national efforts to reduce impacts of fire and other disasters in our communities through education, promoting building codes and standards, fire safety advocacy, data collection, research, and grants—yet there is much more to do.

Millions of Americans witness firsthand how fire continues to pose a substantial risk across the United States. Fire is a public health and safety problem of great proportions, and firefighting remains one of the Nation's most hazardous professions. On average there are more than 1.2 million structure fires, nearly 3,000 deaths, thousands of injuries, and scores of individuals displaced annually from fires. Although disasters such as fires can affect everyone, fires can also exacerbate pre-existing challenges in underserved communities across the country. These impacts are further compounded by poor implementation and enforcement of national building codes and fire risks associated with technology that make fires more common, more intense, and more destructive.

These challenges pose heightened risks to the public and to first responders who safeguard our communities, and the challenge continues to evolve. For example, emerging technologies like Lithium-ion (Li-ion) powered devices and harmful chemicals including polyfluoroalkyl substances (PFAS) introduce new and continued risks to our communities and firefighters.

Emerging Technology

Li-ion batteries are a type of rechargeable battery that contain numerous Li cells. This ordinarily stable electrochemical system provides stored electrical energy, but mechanical, electrical, and thermal abuse and manufacturing defects can destabilize the system and cause thermal runaway. Thermal runaway typically occurs when damaged cells experience uncontrolled increases in temperature and pressure [1]. Thermal runaways can rapidly produce extremely high temperatures in a chain of chemical reactions, and this can induce thermal runaway to propagate to adjacent cells in a battery pack. In addition to heat, Li-ion cells produce flammable gases during thermal runaway that drive Li-ion fires and explosion hazards.

Li-ion batteries are found nearly everywhere. These batteries power everyday items such as cell phones and computers and they are found in e-bikes, e-scooters, and electric vehicles. Li-ion battery energy storage systems are increasingly prevalent at outdoor installations supporting utility operations and installations are expected outside commercial structures and within residences.

While Li-ion batteries are an attractive power option, fire risk increases when they are damaged or used, stored, or charged incorrectly. Combined with what we know of their complex fire risk, their recurring presence requires the fire service to turn research and data into operational considerations quickly.

Li-ion batteries and emerging alternatives constitute a significant component of the drive to reduce emissions worldwide. They are part of a complex global ecosystem of multinational agreements and organizations, geopolitical security questions, and finite natural resources. While a daunting task, the fire service has a central and critical role in ensuring policy decisions address fire safety risks.

Li -ion batteries bring complex operational challenges. Firefighters must consider the presence of Li -ion batteries in all operations, including a risk of faster flashover rates and increased temperatures. Current research shows that Li -ion batteries present four hazard scenarios for firefighters: flammable gas release, flaming, vented deflagrations, and explosions [2].

These batteries are changing the fire risk environment. In a "traditional" fire it typically takes about three minutes or more for a room to be engulfed but now—with the increased prevalence of LI-ion batteries—there is often only 15 seconds from the first sign of smoke to thermal runaway and explosion, with windows being blown out and fire burning in homes, apartments, and businesses. These rapid changes in fire dynamics lead to shorter escape times, shorter time to collapse, and other new and unknown hazards for everyday consumers and for firefighters.

Li-ion batteries also present unusual response challenges. While Li-ion batteries are engineered to be safe, the nature of these devices means they may continue to hold a charge after being damaged, even if fully submerged in water. This phenomenon is known as stranded energy [3].

Therefore, firefighters should always consider whether engineered safety systems are nonfunctional. Li-ion battery fires also can require personnel and water resources far exceeding normal expectations, thereby stressing a department's ability to maintain resources for other emergencies.

While our communities are generally aware of risks associated with their ordinarily benign devices, it is important for the fire service to develop and deploy fire safety messaging regarding safe usage, storage, and charging of Li-ion batteries and their unique risks. As policy decisions are made regarding what can be sold in U.S. markets, the fire service must play a role in discussing the safety of these items, with a specific focus on components directly affecting the fire safety of U.S. communities.

Although research is being conducted to better understand hazards associated with Li-ion batteries and means for mitigation, more research is needed to understand the new and complex hazards Li-ion batteries can present (including exposure to toxic chemicals these batteries release), and to provide firefighters with data and information to inform operational procedures.

Driven by an urgency to meet these risks head-on and save lives, in coordination with the leadership of national fire service organizations, the USFA recently held the second national summit in two years, on fire prevention and control. Together, we assessed the fire problem and challenges faced by firefighters in the United States expanding our 2023 national strategy to address emerging tech. This strategy includes plans to lead and inform discussion on the fire safety of Li-ion batteries and other alternative energy sources within our communities, at all levels of government, and with industry partners utilizing the following priorities:

- Priority 1: Determine risk to the public and firefighter health from lithium-ion battery incidents.
- Priority 2: Conduct a consumer education campaign.
- Priority 3: Conduct necessary research to inform priorities and fire service response organizations in collaboration with our national labs, research institutes, and state and local partners.

PFAS and Firefighter Cancer

PFAS and other toxicants disrupt an individual's fundamental physiology, leading to wide- ranging negative health impacts for firefighters, including cancer and heart disease, as well as sleep and reproductive issues.

Certain PFAS are known to be carcinogenic, and degrade very slowly, earning the label "forever chemicals." PFAS are often found in a firefighter's blood, their firehouses, some firefighting foams, and bunker gear. Next-generation PFAS-free personal protective equipment, along with science and risk-based mitigation programs, can lessen these risks. Firefighters are also exposed to products of combustion, diesel exhaust, building materials, asbestos, chemicals, and ultraviolent radiation.

USFA Actions – Priorities

The 2022 Fire Prevention and Control Summit was the beginning of a comprehensive and strategic approach to addressing impacts of fire on the nation and PFAS as contributing factors to firefighter cancer. The 2023 Summit added fire risks from emerging technology to the National Fire Service Strategy.

As we look to the future, we need data to inform policy and regulation. We must continue our partnerships across the whole-of-fire services, across local, state, and federal governments, our research partners, our non-profit partners, and the industries in the electric vehicle space to determine appropriate regulations to stop deadly tragic events from occurring. Events like those in New York City since 2021where e-bikes with damaged batteries were left to charge overnight and placed in hallways and doors – trapping people inside a burning apartment. Or incidents where people buy aftermarket chargers online – because they are less expensive – leaving battery cells to overcharge, which leads to thermal runaway and fire. New York has pushed forward on regulation of aftermarket chargers because of lessons learned from these incidents. It is our hope that the rest of the nation takes heed and follows suit.

USFA is seeking new ways to address evolving challenges by improving how we collect, analyze, and report relevant information in a timely manner. Legacy National Fire Incident Reporting System (NFIRS) data is inadequate; therefore, USFA is working with the Department of Homeland Security (DHS) Science and Technology Directorate to develop a modern cloud- based data capture system and a streamlined data standard for interoperability and maximum efficiency. The new platform will be known as NERIS (National Emergency Response Information System) and will ensure the USFA and the fire service at-large will have access to secure, interoperable live data that contains outputs from the most authoritative sources. Data scientists and engineers can leverage data from this platform to conduct research and disseminate reports to both the fire service and decision makers at all levels of government.

Additionally, the National Fire Academy (NFA) is increasing its training curriculum to include lithium-ion incident scene safety and fire suppression tactics in existing courses. Major insights were gained from Li-ion batteries found in debris from wildfires in Lahaina on Maui. FEMA and USFA met with the Environmental Protection Agency (EPA) to evaluate a process for de-energizing cells with sustained energy after the fire. Once EPA thoroughly documents the cell de-energization, crushing, and packaging process, the USFA will develop firefighter training curriculum to include this information. NFA curriculum also incorporates training and education for the full spectrum of community risk reduction, and the NFA offers training in multiple mediums to ensure broad access. In Fiscal Year 2023, the NFA delivered training to over 70,000 students.

Regarding PFAS, the "Protecting Firefighters from Adverse Substances Act" (Pub. L. No. 117- 248) (PFAS Act) directs the FEMA Administrator (through USFA) to develop guidance for firefighters and other emergency response personnel on best practices to protect them from exposure to PFAS and to limit and prevent release of PFAS into the environment. Section 2 of the PFAS Act, requires DHS in consultation with the EPA, the Centers for Disease Control and Prevention/National Institute for Occupational Safety & Health (NIOSH), and the heads of other relevant agencies, to:

1. Develop and publish guidance for firefighters on training, education programs and best practices;
2. Make available a curriculum designed to reduce and eliminate exposure, prevent release of PFAS into the environment, and educate firefighters and emergency response personnel on PFAS alternatives; and
3. Create a public repository on tools and best practices to reduce, limit, an prevent the release of and exposure to PFAS.

USFA is actively working on these requirements.

As we look to the challenges ahead, such as those posed by the prevalence of structure fires, the increasing risks of emerging technology, and forever chemicals, USFA looks forward to working with both our firefighting partners and the Members of this Committee to build a fire-safe and more resilient nation. Thank you for the opportunity to testify. I look forward to answering your questions.

Resources

[1] Safety Risks to Emergency Responders from Lithium-Ion Battery Fires in Electric Vehicles. National Transportation Safety Board. November 2020.
[2] The Science of Fire and Explosion Hazards from Lithium-Ion Batteries. UL/Fire Safety Research Institute (FSRI). https://fsri.org/ lithium-ion-battery-guide.
[3] Safety Risks to Emergency Responders from Lithium-Ion Battery Fires in Electric Vehicles. National Transportation Safety Board. November 2020.

Dr. Lori Moore-Merrell, U.S. Fire Administrator

Dr. Lori Moore-Merrell was appointed by President Joseph Biden as the U.S. fire administrator on Oct. 25, 2021. Prior to her appointment, Lori served nearly 3 years as the president and chief executive officer (CEO) of the International Public Safety Data Institute, which she founded after retiring from a 26-year tenure as a senior executive in the International Association of Fire Fighters (IAFF).

Lori is considered an expert in executive leadership, emergency response system evaluation, public safety resource deployment, community risk assessment, data science/analytics, strategic planning, costs and benefits analysis, advocacy, consensus building, policy development and implementation, and generational differences in the workplace.

As a doctor of public health and a data scientist, Dr. Moore-Merrell recently served on the Biden-Harris Transition Team to conduct agency review for the U.S. Department of Homeland Security/Federal Emergency Management Agency (FEMA) as part of the COVID-19 response planning. She has also served on the Public Safety Committee of the transition teams for both the mayor of New York City (2013) and the mayor of the District of Columbia (2015).

Lori serves on national and international boards of directors and advises elected officials, CEOs and local metropolitan fire chiefs in areas of her expertise while providing them scientific data to make fact-based decisions. She was recently awarded Honorary Membership in the Metropolitan Fire Chiefs Association for her expertise in areas of fire prevention, fire suppression or other related disciplines. Lori is only the fourth individual to be presented this honor in the 54-year history of the organization.

Dr. Moore-Merrell is an international speaker, presenter and author. She has also been awarded the James O. Page Achievement Award by the Emergency Medical Services (EMS) Section of the International Association of Fire Chiefs (IAFC) (2001), twice awarded the IAFC President's Award for commitment to firefighter safety (2009, 2019), the Dr.

John Granito Award for Excellence in Fire Leadership and Management Research (2010), the Metropolitan Fire Chiefs President's Award of Distinction (2013), the Mason Lankford Award from the Congressional Fire Services Institute (2019), and the Homeland Security Today Mission Award (2020).

Lori is a principal member of the National Fire Protection Association (NFPA) 3000 Technical Committee and is an advisor to the chair of the NFPA 1710 Technical Committee. Lori served 9 years as a commissioner to the Commission for Fire Service Accreditation and 3 terms as a board member for the Center for Public Safety Excellence. Dr. Moore-Merrell is a member of the International Fire Service Training Association Executive Board, as well as the Underwriters Laboratories (UL) Fire Council.

During her 26 years at the IAFF, Lori spent more than 17 years leading a research team made up of international fire service organizations and other partners, including the National Institute of Standards and Technology, the

UL, the Urban Institute, the University of Texas, Worcester Polytechnic Institute and the Commission on Fire Accreditation International. As principal investigator and senior project manager on projects funded by FEMA/Assistance to Firefighters Grants totaling more than $23 million, she led the team to produce landmark reports and other tools to improve residential and high-rise fireground operations, community risk assessment, fire and EMS resource deployment, and “Big Data Analytics” — all to help drive executive decision-making. These reports and other resources have changed the face of fire and EMS deployment in countries throughout the world.

Dr. Moore-Merrell has also managed emergency response system evaluation project teams, including secure data procurement, geographic information systems analysis and workload analysis in hundreds of fire departments throughout North America.

Lori began her fire service career in 1987 as a fire department paramedic in the City of Memphis Fire Department, Memphis, Tennessee.

Written Statement of Proposed Testimony Chief Fire Marshal Daniel Flynn Fire Department of the City of New York Before the Committee on Homeland Security Subcommittee on Emergency Management and Technology February 15, 2024

Good morning, Chairman D’Esposito, Ranking Member Carter, and the members of the Subcommittee on Emergency Management and Technology.

As Chief Fire Marshal of the New York City Fire Department ("FDNY”), I want to express gratitude to the members of the Subcommittee for holding today’s hearing and to Chairman D’Esposito for inviting me to discuss the dangers of fires involving lithium-ion batteries in micromobility devices. In 2023, New York City experienced 268 fires caused by these batteries in e-bikes, e-scooters, and other micromobility devices. As a result, 150 people were injured and 18 people were killed. These staggering numbers reflect a crisis that has ballooned over a very short period. We’ve seen this problem most acutely in New York. The city has a thriving delivery culture and thousands of delivery workers and messengers who use e-bikes. Many of the deadliest fires have been caused by e-devices being kept in residential homes and apartments. We have begun to see similar issues coast to coast, in communities of all sizes. I speak with counterparts in fire departments across the country and many report the emergence of lithium-ion battery fires and ask for guidance on how to grapple with the issue.

To grasp the urgency of this problem, it's important to understand that fires caused by lithium-ion batteries are more intense and more dangerous than traditional, smoldering fires. Upon ignition, unsafe batteries enter a process called thermal runaway. They undergo a series of explosions, releasing highly toxic gasses, and projecting flaming cells that can travel great distances, increasing the likelihood that the fire will spread. These fires instantly create severelydangerous conditions, rendering escape for anyone nearby significantly challenging. This is especially true if a fire occurs at night when an occupant is sleeping.

Additionally, lithium-ion battery fires require large volumes of water to suppress and can reignite spontaneously, making them extremely difficult for firefighters to extinguish. They also pose uniquely grave dangers for the first responders who respond to these fires and risk their lives every day to protect life and property. One example of the detrimental results of these fires occurred last November, killing three generations of a family in one fire. 81-year-old Albertha West, her son, 58-year-old Michael West, and her grandson, 33-year-old Jamil West perished in that deadly fire because an e-bike containing an uncertified lithium-ion battery erupted in flames. Unfortunately, these fires continue to plague our city and nation.

In my nearly 20-year career, I would be hard-pressed to identify another instance in which a new cause of fires originated and, in only a few years, became one of the leading causes of fatal fires. The FDNY has adjusted quickly, creating new operational procedures for responding, ensuring that the devices are fully under control, and disposing of uncertified batteries and hazardous materials. We created task forces of inspectors who proactively inspect bike shops, respond to complaints, and frequently visit locations most likely to experience problems. We amended department policies, enabling administrative companies to respond immediately to reports of hazardous conditions. We also created robust informational campaigns to educate members of the public about best practices for avoiding problems with their devices. Most importantly, we extensively engaged with our local, state, and federal legislators, seeking support for new laws to help curb the deadly effects of these devices.

At the federal level, New York City respectfully asks Congress to pass H.R. 1797, the Setting Consumer Standards for Lithium-Ion Batteries Act, which would require the Consumer Product Safety Commission to issue a mandatory national standard for these devices. This legislation has bipartisan support and unanimously passed the House Energy and Commerce committee, which is fitting, as these deadly fires do not discriminate: we see them in large cities and small rural areas, in red and blue states alike. Americans need of the protection of Congress, and we hope that this bill is called to an early vote on the House floor.

As we look forward, the experience of lithium-ion batteries serves as a critical reminder of the importance of having public safety entities and first responders at the table when policy is made. Electrification technology is exciting, and there is no shortage of innovators striving to find better solutions. However, it is essential that we implement new technology in concert with an appropriate focus on public safety.

I thank you for your attention to this issue. I am appreciative of the work that you are doing to pass this important legislation, and I know that I share that sentiment with the brave members of the FDNY and grateful firefighters and emergency responders across the country.

Written Statement for the Record

Testimony of Stephen Kerber, PHD, Pe Fire Safety Research Institute, Ul Research Institutes Columbia, Maryland Before the Committee on Homeland Security Subcommittee on Emergency Management and Technology United States House of Representatives For The Hearing Titled "Examining Fire Hazards: Lithium-Ion Batteries and Other Threats to Fire Safety" February 15, 2024

Introduction

Good morning, Chairman D'Esposito, Ranking Member Carter and other members of the subcommittee. I am Steve Kerber, Executive Director of the Fire Safety Research Institute, part of UL Research Institutes. The Fire Safety Research Institute (FSRI) advances fire safety knowledge to address the world's unresolved fire safety risks and emerging dangers. Along with our colleagues in the Electrochemical Safety Research Institute (ESRI), we take on the safety challenges associated with energy technologies. As part of UL Research Institutes, we are committed to sharing our safety insights with everyone to advance UL's public safety mission of providing safe living and working environments for people everywhere. Personally, I have been studying fire safety, with a focus on firefighter health and safety for more than 20 years. I am a 3rd generation volunteer firefighter having served more than a decade in the College Park Fire Department in Prince George's County, Maryland.

Fire Is Fast – And Getting Faster

So far this year, just 6 weeks in (1/1-2/9), we have lost more than 348 people in home fires. At least 50 of those deaths are children, many under the age of five (5). All these deaths are preventable. Americans should be the safest in their homes, but that is simply not the case when it comes to fire safety. Research conducted by FSRI has shown that, during a fire today, you have the least amount of time to safely exit your home than at any time in history.

This is partially because of the synthetic materials used in our furnishings and interior finishes today. It is possible that a fire starting in a bedroom or living room could go from a small flame to flashover – which is when the room becomes fully engulfed with fire – in just three (3) to five (5) minutes. The heat and smoke generated by flashover make conditions unsurvivable in the room where the fire starts and in adjacent rooms or hallways that are open to the fire room. This has contributed to fire deaths steadily increasing over the last decade in the United States. USFA data estimates this increase to be almost 25% since 2012.

The other items, or the fuels, that we bring inside our homes continue to change as well – lithium-ion batteries for example. Lithium-ion batteries have brought essential innovation to our vehicles, our grids, our communities, and everyday products that we rely on. And they are being deployed at a massive scale to drive a reduction in emissions and to improve the resilience of our national electrical grid. But they can overheat, catch fire, and cause explosions with disturbing intensity while emitting toxic smoke. From the first sign of a problem, there could be less than a minute to escape a battery fire.

Fire keeps getting faster as most of the nation's fire departments are ill-equipped to face the threats lithium-ion battery fires pose. That's why we must act now to address escalating fire dangers posed by modern materials and new technologies.

Research is Essential

Despite ongoing safety improvements such as smoke alarms and sprinklers, fires involving lithium-ion battery-powered products are increasing at an alarming rate and have resulted in injuries, fatalities, and property loss. Even when the initial cause of a fire is not the lithium-ion device, the involvement of lithium- ion batteries can increase the intensity and magnitude of any incident. FSRI is honored to support the U.S. Fire Administration and our Fire Service One Voice partners by developing actionable insights through collaborative research on this subject; however, additional research is imperative to reverse the mounting risks presented by this technology.

Experiments and fire investigations have shown that, if damaged or misused, a lithium-ion battery can transition from smoking to explosive fire growth within a matter of seconds. In 2022, a high-profile fire in New York involving lithium-ion batteries injured almost 40 people. The fire was one of hundreds of documented incidents caused by lithium-ion batteries in the United States since 2021. The actual number is likely higher because lithium-ion battery fires are not yet captured by the national fire incident reporting system. These incidents drive the need to better understand the physical phenomena of thermal runaway and the associated hazards.

As Dr. Moore-Merrell described, thermal runaway is one of the primary risks related to lithium-ion batteries. It is a phenomenon in which the lithium-ion cell enters an uncontrollable, self-heating state. We know through our research that thermal runaway can occur undetected until the situation becomes dire and there is an immediate danger of fire. This ultimately translates into shorter escape times and unknown hazards for consumers and first responders.

FSRI is actively investigating multiple facets of battery fires, including:

- Fire service considerations with lithium-ion battery Energy Storage Systems (ESS) [1].
- The potential impacts when lithium-ion battery storage systems fail in homes [2].

- The hazards posed when e-mobility devices (such as e-bikes and scooters) go into thermal runaway [3].
- The fire dynamics and suppression challenges of electric vehicle fires [4].

While this research is already underway, knowledge gaps remain in determining how hazards develop during lithium-ion battery incidents and creating strategies to mitigate the associated risks for Americans and first responders. Further study of materials, construction methods, and computational tools will improve our understanding of the fire dynamics in buildings so that building systems can provide increased protection from lithium-ion battery fires. Experiments focused on lithium-ion battery incidents will characterize risks and advance emergency response protocols. We can't do it alone.

The federal government must collaborate with fire service stakeholders to direct lithium-ion battery research efforts toward the highest priority safety needs. Increased federal funding is imperative to drive research in improving battery safety as technologies advance, ensuring safe functionality of battery systems across electric vehicles, energy storage and other uses of lithium-ion batteries, and equipping fire departments to respond to battery incidents through refined tactics, specialized tools, and reduced chemical exposure risk. Targeted research initiatives and funding in these domains will provide vital progress toward comprehensive lithium-ion battery safety for both the public and first responders.

Safeguarding Fire Responders is Fundamental

Lithium-ion batteries present a dynamic challenge to the fire and emergency services. As use of these devices accelerates through communities, emergency responses will expose first responders to explosive thermal events and toxic emissions beyond traditional protocols. Bravery alone cannot sufficiently protect our fire service. We must have the support of the nation's leaders to ensure the safety of America's communities and fire service personnel. Addressing these challenges will require a multi-faceted process working with a variety of partners (governmental and non-governmental) across many issue areas.

Congress must urgently deliver specialized resources that match the novel threats proliferating. This includes funding:

- Targeted research to equip departments with advanced tactics for battery incidents and integrating the latest science into customized training.
- Modern protective equipment and tools designed specifically to shield against exposure risks distinct to lithium-ion battery chemistry and future chemistries.
- Additional personal protective equipment and tools otherwise out of budgetary reach for resource- starved departments.

It is within your power to direct vital funding so first responders have every chance to prevail over the ever- evolving risks confronting communities across this country. Renewing support for programs like the Assistance to Firefighters Grant (AFG) and Staffing for Adequate Fire and Emergency Response (SAFER) Grant Programs – as well as the U.S. Fire Administration will provide access and accountability nationwide. The safety of Americans begins with securing the safety of our emergency services.

Education is Critical

Both public awareness and first responder training are insufficient regarding lithium-ion battery hazards. Most Americans do not realize the fire risks the ubiquitous devices present if damaged or overcharged. And most fire teams lack the specific protocols needed when battery storage or electric vehicle fire occur.

Leaders must work with fire service stakeholders to prioritize national outreach to address these knowledge gaps. Impactful education includes:

- Mass campaigns conveying battery fire prevention through departments uniquely positioned to connect local constituencies.
- Turnkey campaigns like FRSI's Take C.H.A.R.G.E. of Battery Safety4 initiative designed to promote best practice consumer behaviors and potentially life-saving emergency planning through memorable guidelines.

- Accessible training materials for fire service instruction covering lithium-ion fire dynamics distinct from "traditional" fires and tailored suppression methods that integrate containment, suppression, scene turnover requirements, PPE needs, disposal, and more.

Effective messaging requires distilling cutting-edge research for public comprehension and fire service training customization. We urge officials at all jurisdictional levels to commit resources allowing departments to inform, instruction, and intervene against preventable high-risk battery incidents in the communities they serve. Hazard mitigation begins with awareness.

Governance is Essential

Governance mechanisms around safety standards, trade enforcement measures, and legislative initiatives remain disconnected and outpaced by swiftly evolving technologies. Holistic implementation and enforcement of current codes and standards at federal, state, and local levels will provide the foundation required to properly ensure the fire and life safety ecosystem, especially as codes and standards continue to be updated in response to new and evolving knowledge about technologies, including lithium-ion batteries.

Commercial readiness should never undermine public safeguards. The federal government must look to existing codes and standards organizations like UL Standards & Engagement, the National Fire Protection Association (NFPA), and the International Code Council (ICC) to ensure that laws and regulations consider the work already being done in this space. Applicable codes and standards [5] exist: however, to remain effective, they must constantly integrate manufacturer insights with the latest fire safety research on battery hazards.

The federal government must work with fire service stakeholders to empower consumer regulators with product safety mandates. H.R. 1797/ S.1008, the Setting Consumer Standards for Lithium-Ion Batteries Act, would require the Consumer Product Safety Commission to set a mandatory safety standard for lithium-ion batteries in micro-mobility devices. Closing outdated loopholes that enable uncertified devices into communities is also paramount.

With coordinated governance, the promise of battery technologies can properly and safely accelerate. By infusing regulation with current competence, no innovation outpaces our ability to integrate it responsibly. We urge officials at all levels to partner with researchers and industry in establishing safety protocols.

Conclusion

Public safety must become the top priority as innovations central to our 21st century lifestyles introduce increasing fire hazards. Technologies already promising to transform our vehicles, grids, and communities are now threatening them absent a new paradigm centered on safety.

But with deliberate governance, robust safety standards, and coordinated consumer education, lithium-ion batteries can fulfill their highest purpose responsibly. Implementing the necessary safeguards begins by acknowledging the urgent need for modern research, first responder resources, and public awareness reflective of contemporary risks.

Through immediate action centered on current codes, centralized funding, and community empowerment against preventable risks, we will pass to the next generation safer homes, infrastructure, and the emergency services relied upon to be ever vigilant against hazards both known and still obscure.

Thank you again for the opportunity to share my perspective and I am happy to help this committee address the critical issues it continues to address.

Resources

[1] Fire service considerations with lithium-ion battery ESS (https://training.fsri.org/course/104/fire-service- considerations-with-lithium-ion-battery-energy-storage-systems)

[2] The Impact of Batteries on Fire Dynamics (https://fsri.org/ research/ impact-batteries-fire-dynamics)

[3] Fire Safety Hazards of Lithium-Ion Battery Powered e-Mobility Devices (https://fsri.org/research/examining-fire-safety-hazards-lithium-ion-battery-powered-emobility-devices- homes)

[4] Fire Safety of Batteries and Electric Vehicles (https://fsri.org/research/fire-safety-batteries-and-electric- vehicles)

[5] Take C.H.A.R.G.E. of Battery Safety (https://batteryfiresafety.org; https://vimeo.com/884565314)

[6] For e-mobility, the applicable standards are:

- UL 2054, Household and Commercial Batteries
- UL 2272, Electrical Systems for Personal E-Mobility Devices
- UL 2849, Electrical Systems for eBikes
- UL 2850, Outline of Investigation for Electrical Systems for Electric Scooters and Motorcycles
- UL 2271, Batteries for Use in Light Electric Vehicle (LEV) Applications
- UL 2580, Batteries for Use in Electric Vehicles. Other important standards and codes include:
- UL 9540, Standard for Safety of Energy Storage Systems and Equipment
- NFPA 1, Fire Code
- NFPA 70, National Electrical Code
- NFPA 855, Standard for ESS and Lithium Battery Storage Safety
- International Fire Code, ICC
- International Residential Code, ICC.

Kerber, Stephen			Vice President & Executive Director, Fire Safety
UL Research Institutes			Research Institute
University of Maryland, College Park	BS	2003	Fire Protection Engineering
University of Maryland, College Park	MS	2005	Fire Protection Engineering
Lund University, Lund, Sweden	Ph.D.	2020	Fire Safety Engineering

Licensed Professional Engineer, State of Maryland 36808.

Positions

2019 – Present Vice President, UL Research Institutes, Columbia, MD, USA 2013 – Present Executive Director, Fire Safety Research Institute UL Research Institutes, Columbia, MD, USA

2009 – 2013 Research Engineer Underwriters Laboratories Inc., Northbrook, IL, USA
2002 – 2008 Fire Protection Engineer, National Institute of Standards and Technology, Gaithersburg, MD, USA
1999 – 2009 Deputy Chief College Park Volunteer Fire Department, College Park, MD, USA

Service

Society of Fire Protection Engineers – Fellow, Director, Board of Directors (2012-2017)
National Fire Protection Association - Member, Research Section Executive Committee, Fire Service Training Technical Committee (2008), Fundamentals of Fire Control Within a Structure Utilizing Fire Dynamics (2015-Present), Fire Investigations Committee (2017-Present)
International Association for Fire Safety Science - Member
International Association of Fire Chiefs Regular Member, Volunteer and Combination Officer Section International Society of Fire Service Instructors – Member
University of Maryland Fire Protection Engineering Department - Board of Visitors

Honors

University of Maryland Fire Protection Engineering Distinguished Alumni, 2022 Metropolitan Fire Chiefs President's Award of Distinction 2019
Senator Paul S. Sarbanes Fire Service Safety Leadership Award, UL FSRI 2018 ISFSI George D. Post Fire Service Instructor of the Year 2014
Honorary Battalion Chief, Fire Department of New York 2012 Department of Commerce Gold Medal Award 2008
Research on Wind Driven Fires and Positive Pressure Ventilation Department of Commerce Bronze Medal Award 2007
Research on Positive Pressure Ventilation in High-rise Buildings Department of Commerce Bronze Medal Award 2005
Research on the Station Nightclub Fire

Prince George's County Fire Department Silver Medal of Valor 2002 Fire House Magazine Heroism Award 2002

Selected Peer Reviewed Publications

Kerber, S., Regan, J., Fent K. Horn, G., and D. Smith. "*Effect of Firefighting Intervention on Occupant Tenability during a Residential Fire.*" Fire Technology, 55:2289–2316, 2019.

Kerber, S. "*Analysis of Changing Residential Fire Dynamics and its Implications on Firefighter Operational Timeframes.*" Fire Technology. Volume 48, Number 4, 2012, p 865-891.

Kerber, S. "*Analysis of One and Two-Story Single Family Home Fire Dynamics and the Impact of Firefighter Horizontal Ventilation.*" Fire Technology, Online First 28 August 2012.

Horn, G., Kesler, R., Kerber, S., Fent, K., Schroeder, T., Scott, W., Fehling, P., Fernhall, B. & Smith, D. (2018) Thermal response to firefighting activities in residential structure fires: impact of job assignment and suppression tactic, Ergonomics, 61:3, 404-419.

Kenneth W. Fent, Barbara Alexander, Jennifer Roberts, Shirley Robertson, Christine Toennis, Deborah Sammons, Stephen Bertke, Steve Kerber, Denise Smith & Gavin Horn. (2017) Contamination of firefighter personal protective equipment and skin and the effectiveness of decontamination procedures, Journal of Occupational and Environmental Hygiene, 14:10, 801-814.

Kenneth W. Fent, Douglas E. Evans, Kelsey Babik, Cynthia Striley, Stephen Bertke, Steve Kerber, Denise Smith & Gavin P. Horn (2018) Airborne contaminants during controlled residential fires, Journal of Occupational and Environmental Hygiene, 15:5, 399-412, DOI: 10.1080/15459624.2018.1445260.

Gavin P. Horn, Jacob W. Stewart, Richard M. Kesler, Jacob P. DeBlois, Steve Kerber, Kenneth W. Fent, William S. Scott, Bo Fernhall, Denise L. Smith "Impact of training fire environment on physiological responses" Ergonomics.

Alexander C. Mayer, Kenneth W. Fent, Stephen Bertke, Gavin P. Horn, Denise L. Smith, Steve Kerber, and Mark J. La Guardia "Firefighter Hood Contamination: Laundered vs. Unlaundered" Submitted to JOEH.

Kenneth W. Fent, Christine Toennis, Deborah Sammons, Shirley Robertson, Stephen Bertke, Antonia Calafat, Joachim D. Pleil, M. Ariel Geer

Wallace, Steve Kerber, Denise Smith, Gavin P. Horn. "Firefighters' absorption of PAHs and benzene during controlled residential fires" Submitting to JOEH.

Nicholas Traina, Richard M. Kesler, Steve Kerber, Robin Zevotek, Tonghun Leeb, Gavin P. Horn. "Ex-vivo porcine skin model for prediction of trapped occupant burn risk applicable to pre- and post-suppression fire environments" Submitting to Fire Technology.

Horn, Kesler, Kerber, Fent, Schroeder, Scott, Fehling, Fernhall & Smith (2017): *Thermal response to firefighting activities in residential structure fires: impact of job assignment and suppression tactic*, Ergonomics, DOI: 10.1080/00140139.2017.1355072.

Kerber, S., Madrzykowski, D. *"Wind Driven Structure Fire and Mitigation Strategy Experiments."* 2009 International Symposium on Fire Science and Fire-Protection Engineering Proceedings, Beijing, China, 2009.

Madrzykowski, D., Kerber, S., *"Fire Fighting Tactics Under Wind Driven Conditions: Laboratory Experiments."* National Institute of Standards and Technology, NIST TN 1618, 2009.

Kerber, S., Madrzykowski, *"Evaluating Positive Pressure Ventilation In Large Structures: School Pressure and Fire Experiments."* National Institute of Standards and Technology, NIST TN 1498, 2008.

Kerber, S., *"Evaluation of Fire Service Positive Pressure Ventilation Tactics on Large Structures."*

International Congress: Smoke Control in Buildings and Tunnels Proceedings, Santander, Spain, 2008.

Kerber, S., Madrzykowski, *"Evaluating Positive Pressure Ventilation In Large Structures: High-Rise Fire Experiments."* National Institute of Standards and Technology, NISTIR 7468, 2007.

Kerber, S. *"Evaluation of Fire Service Positive Pressure Ventilation Tactics on High-rise Buildings."* Interflam Proceedings 2007.

Kerber, S., Madrzykowski, D., Stroup, D. *"Evaluating Positive Pressure Ventilation In Large Structures: High-Rise Pressure Experiments."* National Institute of Standards and Technology, NISTIR 7412, 2007.

Kerber, S. *"Evaluation of the Ability of Fire Dynamic Simulator to Simulate Positive Pressure Ventilation in the Laboratory and Practical Scenarios."* National Institute of Standards and Technology, NISTIR 7315, 2006.

Research Support

Previous research support from DHS-FEMA, National Institute of Justice, City of New York, USFA and the US Department of Commerce during tenure at NIST from 2002 – 2008.

Examining Fire Hazards: Lithium-Ion Batteries and Other Threats to Fire Safety
Statement of Fire Chief John S. Butler President and Board Chair, Presented to the Subcommittee on Emergency Management and Technology of the Committee on Homeland Security & Governmental Affairs, United States House of Representatives, February 15, 2024

Good morning, Chairman D'Esposito and Ranking Member Carter. I am John S. Butler, Fire Chief of the Fairfax County (Virginia) Fire and Rescue Department and President and Board Chair of the International Association of Fire Chiefs (IAFC). I appreciate the opportunity today to discuss lithium-ion batteries and other threats to fire safety.

The IAFC represents the leadership of over 1.1 million firefighters and emergency responders. IAFC members are the world's leading experts in firefighting, emergency medical services, terrorism response, hazardous materials (hazmat) incidents, wildland fire suppression, natural disasters, search and rescue, and public-safety policy. Since 1873, the IAFC has provided a forum for its members to exchange ideas, develop best practices, participate in executive training, and discover diverse products and services available to first responders.

America's fire and emergency service is an all-hazards response force that is locally situated, staffed, trained, and equipped to respond to all types of emergencies. There are approximately 1.1 million men and women in the fire and emergency service – consisting of approximately 300,000 career firefighters and 800,000 volunteer firefighters – serving in over 30,000 fire departments around the nation. They are trained to respond to all hazards ranging from earthquakes, hurricanes, tornadoes, and floods to acts of terrorism, hazardous materials incidents, technical rescues, fires, and medical emergencies. We usually are the first at the scene of a disaster and the last to leave.

Dangers Posed by Fires from Lithium-ion Batteries

America's fire and emergency service is approximately five years behind the curve in addressing problems relating to lithium-ion batteries and we request federal assistance to catch up. Fires involving lithium-ion batteries present unique challenges to local fire departments. As a result, local communities must plan for a number of complicated factors. For example, the duration of the fires can be longer: an incident involving an electric vehicle can take four hours and one involving a power storage unit can take approximately 24 hours to extinguish the fire and complete post-fire mitigation. Firefighters must not only extinguish the fire. They also have to pack the device and prepare it for storage to prevent secondary fires. In addition, fire departments also must plan to decontaminate their gear and address concerns about the exposure of firefighters to the toxic smoke caused by a lithium-ion battery fire. All of these characteristics of fires involving lithium-ion batteries can be a burden for fire departments' limited staffing and resources.

Thermal runaway occurs in lithium-ion batteries when the individual cells become destabilized and enter a state of uncontrollable warming. The reaction is the root cause of the fires we see from lithium-ion batteries. Often this phenomenon begins with little to no warning, which can create later complications regarding the removal of any active lithium-ion batteries from an incident scene. Thermal runaway typically presents with large amounts of smoke or gas, which is highly flammable and toxic.

These fires are not just contained to the devices they power. They can engulf the location of the initial fire, along with the surrounding dwellings. The risks posed by lithium-ion fires cannot be understated.

The Source of Fires Involving Lithium-ion Batteries

There is a great chance everyone in this room has some sort of lithium-ion battery on their person. These power the devices which many of us rely upon. Without question, lithium-ion batteries are part of the future of a greener, cleaner society. However, our nation's fire and emergency services have been responding to an increased number of incidents caused by fires involving lithium-ion batteries. Lithium- ion batteries are used to power electric scooters; electric bikes; hoverboards; wheelchairs; personal computers; cell phones; landscaping tools; electronic cigarettes; golf carts; energy storage systems used to power homes; all-terrain vehicles; electric vehicles (EV); commercial buses; trucks; and much more.

Our fire service is not just challenged in the EV and mobility space with completed products. With the rapid increase of cell manufacturing in the United States, many of our communities are struggling with new facilities that are part of the rapidly growing manufacturing sector. This may be a battery plant; manufacturing facility; automobile assembly; or even a battery laboratory. Our firefighters are challenged with not just the buildings, but the transit of these materials as part of the complete ecosystem. This can include the safe transport of products by rail, road, and waterways.

The battery is not the lone concern with the operation of many of these devices. The charging components, use of third-party replacement lithium-ion batteries, and large home energy storage systems also pose great concerns. It is also worth noting the growing prevalence of home energy storage systems that use lithium-ion batteries to power an entire home. Unfortunately, the increase of these batteries in our society has not led to increased response capabilities for the fire service or rapid adoption of current and model fire and building codes. The fire service is at the initial stages of exploring the best and safest methods to respond to fires caused by these batteries. Several things need to happen to make these devices safe for all. We need more help.

Lithium-ion Battery Fires Are a National Problem

I would like to thank my partners at the FDNY for their leadership in this effort. Without a doubt, New York City has experienced a high number of these fires. Over the last four years in New York alone, there were more than 400 fires related to lithium-ion batteries. These fires resulted in more than 300 injuries, 12 deaths and damage to more than 320 structures and more than 100 non-structures. As a response to these fires, the FDNY is one of the most proactive voices calling for the necessary enactment of laws and regulations to try and remedy this situation.

Nonetheless, I would like to call attention to how lithium-ion battery fires are affecting communities all over our nation.

- In Fairfax County (VA), we had 17 incidents involving lithium-ion batteries in 2023. They were in a variety of devices including vehicles, mobile phones, portable chargers, laptop computers, and remote-controlled cars.
- In March 2021, the Harrisburg (PA) Bureau of Fire experienced a line of duty death due to a fire caused by lithium-ion batteries found in hoverboards.
- In March 2023, the Brighton (MI) Area Fire Authority experienced three fires in one week, which involved a plug-in hybrid, a cell phone battery; and a mobility-based device.
- On March 2023, a lithium-ion battery from a hoverboard ignited a basement fire in Lodi, NJ.
- During 2023, Houston (TX) experienced more than 60 fires involving rechargeable lithium-ion batteries. These fires included lithium-ion batteries in hoverboards, scooters, and motor vehicles.
- Gainesville (FL) experienced several fires due to devices powered by lithium-ion batteries. In 2023, two of these fires involved surrounding structures and dwellings.

Washington D.C. experienced eight fires in 2023 that were attributed to lithium-ion batteries. Three of these fires involved e-bikes and scooters and one involved a hoverboard.

The Need for Better Data on Lithium-Ion Fires

As the nation deals with an increase in lithium-ion battery fire incidents, it is important that we can track and better understand their occurrences. We need to know the answers to questions like "What devices cause these incidents? Who are the operators of these devices? Where are these incidents occurring, and how often?" These metrics will help us better understand how and where to allocate resources. Some representatives of the fire service are tracking this data and some states, like Florida, are beginning to require their fire departments to report lithium-ion battery fires. Yet, we still need a national understanding of the scope of the problem of lithium-ion fires.

Currently, the national fire and emergency service utilizes the National Fire Incident Reporting Systems (NFIRS) to track fire-related incidents. The United States Fire Administration (USFA) is developing a replacement for NFIRS that will include real-time data on fires. The IAFC supports the USFA's effort to develop the National Emergency Response Information System (NERIS) and urges Congress to fully fund its development. The development of NERIS will give our communities the necessary tools to track information about incidents involving lithium-ion batteries. With a better understanding of the scope of the problem, Congress and the Administration will be able to allocate resources to help local fire departments respond to this growing problem.

The Need to Develop Codes and Standards

To prohibit the further entry of faulty lithium-ion batteries into our communities, model codes and standards must be developed, updated, and adopted. This will involve collaboration between many stakeholders, such as the fire service; federal, state, and local agencies; research organizations; and manufacturers. There are several ways this can be achieved:

1. Look to notable fire service organizations that are leaders in the code and standard space, such as Underwriters Laboratories (UL), the National Fire Protection Association (NFPA), and the International Code Council (ICC). The fire and emergency service needs strong partners that not only work to reduce these events from happening, but encourage the industry to develop solutions for post-incident mitigation of a fire where lithium-ion batteries are involved.

Organizations such as UL, NFPA, and ICC are leading the effort to adapt codes and standards to adapt to technology using lithium-ion batteries. Further support of their work will lead to increased safety for not just consumers, but also first responders, who respond to lithium-ion battery fires. While we all know the power of using modern building and fire codes, states and communities also need support in adopting the most current codes and standards to address this rapidly changing industry.

2. Pass and enact the Setting Consumer Standards for Lithium-Ion Batteries Act (H.R. 1797/ S. 1009). This legislation would require the Consumer Product Safety Commission to issue safety standards on lithium-ion batteries in mobility devices. A high percentage of the fires caused by lithium-ion batteries are in devices like e-bikes, e-scooters, and hoverboards. With the increased use of micro-mobility devices powered by lithium-ion batteries, it is paramount that we set safety standards to ensure that consumers are not subject to harm. H.R. 1797, as amended, advanced out of the House Energy and Commerce Committee by a total of 42-0. I urge the full House of Representatives to swiftly consider this legislation and pass it without delay. In the meantime, I urge the Senate Committee on Commerce, Science, and Transportation Committee to begin consideration of S. 1009. The sooner Congress acts, the faster we can start to prevent unsafe lithium-ion batteries from being on America's streets and in American homes.

The Need for More Training, Resources, and Increased Public Education

Fire departments should work with organizations like the UL's Fire Safety Research Institute, the New York City Fire Department (FDNY), the USFA, and the IAFC to prepare for lithium-ion fires. The USFA can use the National Fire Academy and its relationship with the state and local fire training academies to train firefighters about how to respond to fires caused by lithium-ion batteries and how to safely manage and cleanup the incident scene after the fire. In addition, the USFA can develop public education campaigns to educate the public about the safe handling and storage of devices and vehicles using lithium-ion batteries.

The Assistance to Firefighters Grant (AFG) and the Staffing for Adequate Fire and Emergency Response (SAFER) grants provide matching grants that can be used to help local fire departments with incidents involving lithium-ion batteries. However, both programs already cannot meet the current demand for their funds. In addition, unless Congress passes the Fire Grants and Safety Act (H.R. 4090/S. 870), these programs will expire on September 30, 2024.

We recommend that Congress develop a new program that will help communities prepare for incidents involving lithium-ion batteries. The program should fund code adoption efforts; planning, training, and exercises; and equipment. The program also should fund research into issues like firefighter exposure to toxic fumes from lithium-ion fires and how to effectively decontaminate gear that has been used in a fire involving lithium-ion batteries.

Conclusion

I thank you for the opportunity to address the threat of lithium-ion batteries and other threats to fire safety. While lithium-ion batteries present a promise in providing power to new forms of technology, we must take steps to prepare for accidental fires caused by them. Congress can play a role in ensuring the nation's preparedness by passing legislation like the Setting Consumer Standards for Lithium-Ion Batteries Act (H.R. 1797/ S. 1009). In addition, it can pass the Fire Grants and Safety Act (H.R. 4090/ S. 870) to preserve programs like the AFG and SAFER programs and also create a program to help local communities work with their fire departments to prepare for incidents involving lithium-ion batteries. We also support increased funding for the USFA to ensure that it can improve data collection efforts and distribute training and public education to help local communities prevent fires involving lithium-ion batteries. The IAFC looks forward to working with the committee to ensure the safe adoption of this revolutionary new technology.

Chapter 3

Setting Consumer Standards for Lithium-Ion Batteries Act*

Committee on Energy and Commerce

The Committee on Energy and Commerce, to whom was referred the bill (H.R. 1797) to require the Consumer Product Safety Commission to promulgate a consumer product safety standard with respect to rechargeable lithium-ion batteries used in micromobility devices, and for other purposes, having considered the same, reports favorably thereon with an amendment and recommends that the bill as amended do pass.

SECTION 1. SHORT TITLE.

This Act may be cited as the "Setting Consumer Standards for Lithium-Ion Batteries Act".

SEC. 2. CONSUMER PRODUCT SAFETY STANDARD FOR CERTAIN BATTERIES.

(a) CONSUMER PRODUCT SAFETY STANDARD REQUIRED.—

(1) IN GENERAL.—Not later than 1 year after the date of the enactment of this Act, the Consumer Product Safety Commission shall promulgate, under section 553 of title 5, United States Code, a final consumer product safety standard for rechargeable lithium-ion batteries used in micromobility devices, including electric bicycles and electric scooters, to protect against the risk of fires caused by such batteries.

(2) INCLUSION OF RELATED EQUIPMENT.—The standard promulgated under paragraph (1) shall include requirements with respect to equipment related to or used with rechargeable lithiuion batteries used in

* This is an edited, reformatted and augmented version of Report Committed to the Committee of the Whole House on the State of the Union and Ordered to Be Printed to accompany H.R. 1797 Including cost estimate of the Congressional Budget Office, dated April 5, 2024.

In: Lithium-Ion Batteries
Editor: Gregory R. Simmons
ISBN: 979-8-89530-773-1

micromobility devices, including battery chargers, charging cables, external terminals on battery packs, external terminals on micromobility devices, and free-standing stations used for recharging.

(b) CPSC DETERMINATION OF SCOPE. — In promulgating the standard under subsection (a), the Commission shall determine the types of products subject to the standard and shall ensure that such products are—

(1) within the jurisdiction of the Commission; and

(2) reasonably necessary to include to protect against the risk of fires.

(c) MODIFICATIONS. — At any time after the promulgation of the standard under subsection (a), the Commission may, through a rulemaking under section 553 of title 5, United States Code, modify the requirements of the standard.

(d) TREATMENT OF STANDARD. — A standard promulgated under this section, including a modification of such standard, shall be treated as a consumer product safety rule promulgated under section 9 of the Consumer Product Safety Act (15 U.S.C. 2058).

Purpose and Summary

H.R. 1797, the "Setting Consumer Standards for Lithium-Ion Batteries Act," was introduced by Representative Torres on March 24, 2023, and was referred to the Committee on Energy and Commerce. H.R. 1797 requires the Consumer Product Safety Commission (CPSC) to promulgate a consumer product safety standard to protect consumers from the risk of fires associated with rechargeable lithium-ion batteries used in micromobility devices.

Background and Need for Legislation

Lithium-Ion batteries are lightweight, rechargeable batteries found in many consumer electronics and are often used in micromobility devices, such as electric bikes and scooters. When poorly made, lacking adequate safety testing, charged improperly, or damaged these batteries are prone to ignite and the associated fires may be accompanied by explosions and the release of toxic gas.[100] As micromobility devices have risen in popularity, the use of

[100] National Fire Protection Association, *Lithium-Ion Battery Safety* (accessed Jan. 4, 2024) (https://www.nfpa.org/education-and-research/home-fire-safety/lithium-ion-batteries).

lithiumion batteries has increased, creating the need for safety standards. Currently, there is no federal safety standard for Lithium-Ion batteries[101] and many uncertified and untested batteries are available for purchase.[102]

From 2019 to 2023, the Fire Department of New York reported more than 400 fires, 300 injuries, and twelve deaths caused by lithium-ion batteries in New York City alone.[103] Urban areas are at increased risk for injuries and property damage due to high population density, but Lithium-ion battery fires impact communities across the United States. Consumer advocates and fire professionals have warned consumers only to use certified and tested products and called for strong federal safety standards.[104]

Committee Action

On September 27, 2023, the Subcommittee on Innovation, Data, and Commerce held a hearing on H.R. 1797. The title of the hearing was "Proposals to Enhance Product Safety and Transparency for Americans." The Subcommittee received testimony from:

- Kathleen Callahan, Owner, Xpertech Auto Repair;
- Scott Benavidez, Chairman, Automotive Service Association;
- Steven Michael Gentine, Counsel, Arnold & Porter, LLP;
- John Breyault, Vice President of Public Policy, Telecommunications and Fraud, National Consumers League; and,
- David Touhey, Principal, Connett Consulting, appearing on behalf of International Association of Venue Managers.

On November 2, 2023, the Subcommittee on Innovation, Data, and Commerce met in open markup session and forwarded H.R. 1797, as amended, to the full Committee by a record vote of 20 yeas and 0 nays.

On December 6, 2023, the full Committee on Energy and Commerce met in open markup session and ordered H.R. 1797, without amendment, favorably reported to the House by a record vote of 42 yeas and 0 nays.

---

[101] Letter from International Association of Fire Fighters, to Subcommittee on Innovation, Data, and Commerce Chair Gus Bilirakis and Ranking Member Jan Schakowsky (Sept. 26, 2023).

[102] International Association of Fire Fighters, *Preventing Lithium-Ion Battery Fires*, (July 18, 2023) (https://www.iaff.org/news/preventing-lithium-ion-battery-fires/).

[103] See Note 2.

[104] See Note 1.

Committee Votes

Clause 3(b) of rule XIII requires the Committee to list the record votes on the motion to report legislation and amendments thereto. The following reflects the record votes taken during the Committee consideration: -

COMMITTEE ON ENERGY AND COMMERCE
118TH CONGRESS
ROLL CALL VOTE # 9

BILL: H.R. 1797, Setting Consumer Standards for Lithium-Ion Batteries Act

AMENDMENT: A motion by Chair Rodgers to order H.R. 1797 favorably reported to the House, as amended (Final Passage).

DISPOSITION: **AGREED TO,** by a roll call vote of 42 yeas and 0 nays.

REPRESENTATIVE	YEAS	NAYS	PRESENT	REPRESENTATIVE	YEAS	NAYS	PRESENT
Rep. Rodgers	X			Rep. Pallone	X		
Rep. Burgess	X			Rep. Eshoo	X		
Rep. Latta	X			Rep. DeGette			
Rep. Guthrie	X			Rep. Schakowsky	X		
Rep. Griffith	X			Rep. Matsui			
Rep. Bilirakis	X			Rep. Castor	X		
Rep. Johnson	X			Rep. Sarbanes	X		
Rep. Bucshon	X			Rep. Tonko	X		
Rep. Hudson	X			Rep. Clarke	X		
Rep. Walberg	X			Rep. Cárdenas	X		
Rep. Carter	X			Rep. Ruiz	X		
Rep. Duncan	X			Rep. Peters	X		
Rep. Palmer	X			Rep. Dingell			
Rep. Dunn				Rep. Veasey	X		
Rep. Curtis				Rep. Kuster	X		
Rep. Lesko	X			Rep. Kelly	X		
Rep. Pence	X			Rep. Barragán	X		
Rep. Crenshaw				Rep. Blunt Rochester	X		
Rep. Joyce	X			Rep. Soto	X		
Rep. Armstrong				Rep. Craig			
Rep. Weber	X			Rep. Schrier	X		
Rep. Allen	X			Rep. Trahan	X		
Rep. Balderson	X			Rep. Fletcher			
Rep. Fulcher							
Rep. Pfluger	X						
Rep. Harshbarger	X						
Rep. Miller-Meeks	X						
Rep. Cammack	X						
Rep. Obernolte	X						

12/06/2023

Oversight Findings and Recommendations

Pursuant to clause 2(b)(1) of rule X and clause 3(c)(1) of rule XIII, the Committee held a hearing and made findings that are reflected in this report.

New Budget Authority, Entitlement Authority, and Tax Expenditures

Pursuant to clause 3(c)(2) of rule XIII, the Committee finds that H.R. 1797 would result in no new or increased budget authority, entitlement authority, or tax expenditures or revenues.

Congressional Budget Office Estimate

Pursuant to clause 3(c)(3) of rule XIII, the following is the cost estimate provided by the Congressional Budget Office pursuant to section 402 of the Congressional Budget Act of 1974:

At a Glance

Health Care Legislation

As ordered reported by the House Committee on Energy and Commerce on December 6, 2023

On December 6, 2023, the House Committee on Energy and Commerce ordered reported 41 bills related to health care and energy. This single, comprehensive document provides estimates for 21 bills related to health care and consumer protection.

Five bills would affect spending subject to appropriation. Ten would affect direct spending; thus, pay-as-you-go procedures apply. One bill would significantly increase net direct spending or on-budget deficits in at least one of the four consecutive 10-year periods beginning in 2035. Three bills would impose private-sector mandates. Details of the estimated costs of each bill are discussed in the text below.

Bill	Net Increase or Decrease (¥) in the Deficit Over the 2024–2034 Period (Millions of Dollars)	Changes in Spending Subject to Appropriation Over the 2024–2029 Period (Outlays, Mil- lions of Dollars)	Mandate Effects?
H.R. 133	0	0	No
H.R. 1797	0	6	Yes
H.R. 2365	0	3	No
H.R. 2880	¥226	0	No
H.R. 3842	*	0	No
H.R. 4310	0	2	Yes

Bill	Net Increase or Decrease (¥) in the Deficit Over the 2024–2034 Period (Millions of Dollars)	Changes in Spending Subject to Appropriation Over the 2024–2029 Period (Outlays, Mil- lions of Dollars)	Mandate Effects?
H.R. 4881[a]	754	0	No
H.R. 5202	0	22	No
H.R. 5371	0	0	No
H.R. 5372	¥145	0	No
H.R. 5380	15	0	No
H.R. 5385	¥381	0	No
H.R. 5386	*	0	No
H.R. 5388	0	0	No
H.R. 5389	0	0	No
H.R. 5393	6	0	No
H.R. 5396	0	0	No
H.R. 5397	¥139	0	No
H.R. 5555	145	0	No
H.R. 6132	0	3	Yes
H.R. 6364	0	0	No

* = between ¥$500,000 and $500,000.

[a] H.R. 4881 would increase on-budget deficits by more than $5 billion in at least one of the four consecutive 10-year periods beginning in 2035.

Summary: On December 6, 2023, the House Committee on Energy and Commerce ordered 41 pieces of legislation to be reported. This document provides estimates for 21 bills in that package that are related to health care and consumer protection.

Generally, the bills in this group that would affect direct spending would:

- Limit beneficiary cost sharing for certain prescription drugs and add certain drugs to the group of products covered by the Medicare home infusion benefit;
- Prohibit pharmacy benefit managers (PBMs) from collecting certain fees from prescription drug manufacturers and require PBMs to provide additional information to Medicare Part D plans (which provide prescription drug coverage);
- Allow Part D plans more flexibility to add biosimilar biological products to their formularies and to change the cost- sharing status of reference biological products;
- Temporarily increase Medicare payment rates for durable medical equipment (DME); and

- Provide mandatory funding for implementation of certain provisions in several bills.

Estimated Federal cost: The costs of the legislation fall within budget functions 550 (health) and 570 (Medicare).

Table 1. Estimated Effects on Direct Spending of Health Care Legislation, as Ordered Reported by the House Committee on Energy and Commerce on December 6, 2023

	By fiscal year, millions of dollars—												
	2024	2025	2026	2027	2028	2029	2030	2031	2032	2033	2034	2024–2029	2024–2034
	Increases or Decreases (−) in Direct Spending												
H.R. 2880:													
Budget Authority	0	0	0	−29	−39	−31	−31	−28	−26	−24	−18	−99	−226
Outlays	0	0	0	−29	−39	−31	−31	−28	−26	−24	−18	−99	−226
H.R. 3842:													
Budget Authority	0	*	*	*	*	*	*	*	*	*	*	*	*
Outlays	0	*	*	*	*	*	*	*	*	*	*	*	*
H.R. 4881:													
Budget Authority	0	0	0	0	77	88	100	106	113	134	136	165	754
Outlays	0	0	0	0	77	88	100	106	113	134	136	165	754
H.R. 5372:													
Budget Authority	0	−9	−12	−12	−14	−12	−14	−16	−17	−20	−19	−59	−145
Outlays	0	−9	−12	−12	−14	−12	−14	−16	−17	−20	−19	−59	−145
H.R. 5380:													
Budget Authority	15	0	0	0	0	0	0	0	0	0	0	15	15
Outlays	13	1	1	0	0	0	0	0	0	0	0	15	15
H.R. 5385:													
Budget Authority	55	0	0	−55	−75	−60	−60	−55	−50	−46	−35	−135	−381
Outlays	49	4	2	−55	−75	−60	−60	−55	−50	−46	−35	−135	−381
H.R. 5386:													
Budget Authority	0	0	*	*	*	*	*	*	*	*	*	*	*
Outlays	0	0	*	*	*	*	*	*	*	*	*	*	*
H.R. 5393:													
Budget Authority	0	6	0	0	0	0	0	0	0	0	0	6	6
Outlays	0	6	0	0	0	0	0	0	0	0	0	6	6
H.R. 5397:													
Budget Authority	0	0	−9	−13	−15	−14	−15	−17	−17	−20	−19	−37	−139
Outlays	0	0	−9	−13	−15	−14	−15	−17	−17	−20	−19	−37	−139
H.R. 5555:													
Budget Authority	144	1	0	0	0	0	0	0	0	0	0	145	145
Outlays	144	1	0	0	0	0	0	0	0	0	0	145	145

All amounts for outlays are estimates; except for H.R. 5380 and H.R. 5393, all amounts for budget authority are estimated; * = between ¥$500,000 and $500,000.

Basis of estimate: For this estimate, CBO assumes that the bills will be enacted near the middle of fiscal year 2024 and that the estimated amounts will be appropriated each year. This cost estimate does not include any effects of interactions among the bills. If all 21 bills were combined and enacted as a single piece of legislation, the effects could be different from the sum of the separate estimates.

Direct spending: Enacting 10 bills in the group would affect direct spending over the 2024–2034 period (see Table 1).

H.R. 2880, the Protecting Patients Against PBM Abuses Act, would modify the rules with respect to certain fees that PBMs collect from prescription drug manufacturers. In Medicare Part D (which provides prescription drug coverage), sponsors of private insurance plans contract with the government to deliver benefits to Medicare beneficiaries. Those insurance plans usually contract with PBMs to negotiate with drug manufacturers, design formularies, and perform other administrative functions. A PBM can be owned by the plan sponsor or it can be an independent corporate entity.

H.R. 2880 would prohibit PBMs from collecting service fees from manufacturers that are based on drug prices, manufacturer discounts, or formulary placement decisions. Under the bill, those fees would be specific dollar amounts based on the fair market value of a PBM's services. Under current law, PBMs can be compensated for services they provide to manufacturers, but compensation that exceeds the fair market value of a service must be classified as direct and indirect remuneration and reported to the Centers for Medicare & Medicaid Services (CMS). According to the Government Accountability Office, however, CMS does not routinely monitor how PBMs classify those fees.[105] Under the bill, CMS and the Office of Inspector General would more closely monitor those classifications.

CBO estimates that manufacturers' service fees are roughly 1 percent of Part D retail spending under current law. CBO expects that under H.R. 2880, a portion of those fees would be reclassified as direct and indirect remuneration by PBMs and, because of stronger oversight, passed along to the sponsors of prescription drug plans. That action would reduce bid amounts for plans' expected benefit payments, which in turn would reduce spending in Part D. CBO estimates that the provision would decrease federal spending by $226 million over the 2024–2034 period, or by roughly 1 percent of the amount expected to be collected in service fees over that period.

H.R. 3842, the Expanding Access to Diabetes Self-Management Training Act of 2023, would allow more providers to refer eligible patients to diabetes self-management training covered by Medicare and would codify regulatory

---

[105] See Government Accountability Office, *Medicare Part D: Use of Pharmacy Benefit Managers and Efforts to Manage Drug Expenditures,* GAO–19–498 (July 2019), Appendix III, www.gao.gov/ products/gao-19-498.

time limits on use of the training. CBO expects that enacting H.R. 3842 would result in more patients receiving such training, which would lead to increased Medicare spending. CBO expects that such training would reduce the use of acute-care services, at least partly offsetting that increase in costs. As a result, CBO estimates that enacting the bill would increase or decrease direct spending by less than $500,000 over the 2024– 2034 period.

H.R. 4881, a bill to amend title XVIII of the Social Security Act to limit cost sharing for drugs under the Medicare program, would limit cost sharing above the deductible to no more than the average net price for a drug, which is the list price minus after-sale discounts from the drug's manufacturer. From 2028 to 2034, CBO projects, less than 1 percent of Part D spending above the deductible under current law will be for drugs with cost sharing that exceeds net drug costs. Under the bill, CBO expects that some out- of-pocket spending by beneficiaries and some federal subsidies for low-income beneficiaries would shift onto Part D plans, which would increase the bids they submit to the federal government to cover expected benefits spending and therefore increase federal spending. CBO estimates that enacting H.R. 4881 would increase direct spending by $754 million over the 2024–2034 period.

H.R. 5372, the Expanding Seniors' Access to Lower Cost Medications Act of 2023, would allow Part D plans to add biosimilar biological products to their formularies and change the cost-sharing status of a reference biological product after the first 60 days of a plan year. (A reference biological product is the approved product against which a proposed biosimilar product is compared.) Under current law, Part D plans must exempt beneficiaries who currently use reference biological products from changes in coverage and cost sharing for the remainder of the year. That restriction limits a plan's ability to promote use of a biosimilar product immediately following that product's entry to the market. CMS has proposed rules that overlap with the bill's provisions concerning formulary substitutions for biosimilar products.[106]

[106] See Centers for Medicare & Medicaid Services, ''Medicare Program; Contract Year 2025 Pol- icy and Technical Changes to the Medicare Advantage Program, Medicare Prescription Drug Benefit Program, Medicare Cost Plan Program, and Programs of All-Inclusive Care for the Elderly; Health Information Technology Standards and Implementation Specifications,'' Notice of Proposed Rulemaking, 88 *Fed. Reg.* 78476 (November 15, 2023), http://tinyurl.com/wv7yprfm; and ''Medicare Program; Contract Year 2024 Policy and Technical Changes to the Medicare Advantage Program, Medicare Prescription Drug Program, Medicare Cost Plan Program, Medicare Parts A, B, C, and D Overpayment Provisions of the Affordable Care Act and Programs of All- Inclusive Care for the Elderly; Health Information Technology Standards and Implementation Specifications,'' Notice of Proposed Rulemaking, 87 *Fed. Reg.* 79452 (December 27, 2022), http:// tinyurl.com/3754c49x.

CBO's estimate of Medicare spending for those products under current law accounts for 50 percent of the effect of the proposed rules. As a result, CBO's estimate of the decrease in direct spending under H.R. 5372 is larger than it might be if CMS's rules had become final.

Under the bill, the addition of biosimilar products to formularies could lead to a shift away from the use of reference biological products. CBO estimates that the government will spend about $10 billion over the 2024–2034 period to cover reference biological products under current law. CBO anticipates that under H.R. 5372 approximately 20 percent of the current use of reference biological products would be replaced by biosimilar products. The prices for biosimilar products are estimated to be 15 percent lower, on average, than the prices for the reference products. Using information about spending on both types of products under current law and adjusting for current regulatory proposals by CMS that would streamline coverage for biosimilar products, CBO estimates that enacting H.R. 5372 would decrease direct spending by $145 million over the 2024–2034 period.

H.R. 5380, a bill to amend title XVIII of the Social Security Act to increase data transparency for supplemental benefits under Medicare Advantage, would provide $15 million in 2024 for the Department of Health and Human Services (HHS) to implement reporting requirements for supplemental benefits under Medicare Advantage plans. Based on historical spending patterns for HHS programs, CBO estimates that enacting H.R. 5380 would increase direct spending by $15 million over the 2024–2034 period.

H.R. 5385, the Medicare PBM Accountability Act, would require pharmacy benefit managers to provide plan sponsors with information not furnished under current law. Part D plans have access to certain aggregate and drug-specific information from PBMs concerning prescriptions, prices, rebates, and out-of-pocket charges, but may lack information about PBM-affiliated entities and contractors, rationales for formulary decisions, and explanations for benefit designs that favor certain pharmacies. H.R. 5385 would require PBMs to report such information to Part D plans but also, subject to certain restrictions, would allow plans to audit PBMs' business practices and request other information. The bill would provide $55 million for HHS to implement those requirements.

H.R. 5385 also would require PBMs to make their business practices clearer to Part D plans, thus promoting competition among PBMs. CBO estimates that the increased competition would reduce net spending for Part D by less than 0.1 percent over the 2024– 2034 period—reducing federal spending by $436 million over that period.

CBO estimates that the net effect of the bill would be a reduction in direct spending of $381 million over the 2024–2034 period.

H.R. 5386, the Cutting Copays Act, would prohibit cost sharing for generic drugs for beneficiaries who are eligible for the low-in- come subsidy, which pays most or all of their premium and cost- sharing requirements. Under current law, plans have an option but not an obligation to do so. CBO expects that enacting the bill would increase the use of generic drugs, which would increase plan bid submissions for expected benefits payments and, therefore, federal spending. CBO expects that some of the increase would be offset by reduced spending on brand-name drugs and certain medical services. CBO estimates that enacting the bill would increase direct spending by less than $500,000 over the 2024–2034 period.

H.R. 5393, a bill to amend title XVIII of the Social Security Act to ensure fair assessment of pharmacy performance and quality under Medicare Part D, and for other purposes, would provide $4 million in 2025 for CMS program management to implement pharmacy performance and quality measures for Part D and $2 million in that year to implement pharmacy transparency requirements. Based on historical spending patterns for CMS administrative costs, CBO estimates that enacting H.R. 5393 would increase direct spending by $6 million over the 2024–2034 period.

H.R. 5397, the Joe Fiandra Access to Home Infusion Act of 2023, would add drugs to the current Medicare benefit that allows patients to receive some drugs by infusion under nursing care at home. H.R. 5397 would allow other drugs to meet the statutory criteria for coverage in the home setting by establishing those products as suitable for delivery through a pump and requiring patients receiving those drugs also to receive regular nursing services.

Based on its analysis of the beneficiary population and Medicare payment rates, CBO estimates that enacting the bill would reduce direct spending by $139 million over the 2024–2034 period, primarily because beneficiaries would bear a larger share of the cost of infusions that occur at home. Under current law, there is a cap on beneficiary cost sharing in outpatient hospital settings, which is where CBO expects that beneficiaries receive those drugs now. There is no equivalent cap for the home infusion benefit.

CBO's estimate for H.R. 5397 is subject to considerable uncertainty. First, it is not known how many drugs would qualify for coverage under the bill. CBO's estimate focused on three products that industry and clinical experts mentioned as likely candidates, but the actual number could be larger or smaller. In addition, given that cost sharing could increase significantly for patients, it is not known how many beneficiaries would choose to receive home infusions.[107]

H.R. 5555, the DMEPOS Relief Act of 2023, would temporarily increase Medicare rates in some areas of the country for DMEPOS (durable medical equipment, prosthetics, orthotics, and supplies). Under current law, Medicare's payments for some equipment are based on competitive bidding among suppliers. CMS uses those results to set rates (either directly or through a blend with the historic fee schedule) in areas of the country where formal bidding has not occurred. Prior legislation directed CMS to use a blend of fee schedule and competitively bid rates in some areas of the country; the use of those blended rates expired at the end of calendar year 2023. Enacting H.R. 5555 would extend the use of those blended rates through calendar year 2024. Based on an analysis of historic claim spending, CBO estimates that the DME provision of the bill would increase direct spending by $145 million over the 2024–2034 period. H.R. 5555 also would reduce amounts available to the Medicare Improvement Fund by $177 million, however the Consolidated Appropriations Act, 2024 rescinded all funding from the Medicare Improvement Fund. As a result, the provision would not affect direct spending. In total, CBO estimates that enacting H.R. 5555 would increase net direct spending by $145 million over the 2024– 2034 period.

Legislation with no effect on direct spending: CBO estimates that enacting 11 bills in this estimate would have no effect on direct spending over the 2024–2034 period:

- H.R. 133, the Mandating Exclusive Review of Individual Treatments (MERIT) Act;
- H.R. 1797, the Setting Consumer Standards for Lithium- Ion Batteries Act;
- H.R. 2365, the Dr. Emmanuel Bilirakis National Plan to End Parkinson's Act;
- H.R. 4310, the Youth Poisoning Protection Act;

[107] CMS proposed a similar but not identical policy in a proposed rulemaking. In the regulatory impact analysis, CMS estimated that, for one product, beneficiaries' cost sharing would be about triple the amount if the product was received in a home setting. For more information, see Centers for Medicare & Medicaid Services, ''Medicare Program; Durable Medical Equipment, Prosthetics, Orthotics, and Supplies (DMEPOS) Policy Issues and Level II of the Healthcare Common Procedure Coding System (HCPCS),'' Notice of Proposed Rulemaking, 85 Fed. Reg. 70358 (November 4, 2020), http://tinyurl.com/29djdrvz.

- H.R. 5202, the Virginia Graeme Baker Pool and Spa Safety Reauthorization Act;
- H.R. 5371, the Choices for Increased Mobility Act of 2023;
- H.R. 5388, the Supporting Innovation for Seniors Act;
- H.R. 5389, the National Coverage Determination Transparency Act;
- H.R. 5396, the Coverage Determination Clarity Act of 2023;
- H.R. 6132, the Awning Safety Act of 2023; and
- H.R. 6364, the Medicare Telehealth Privacy Act of 2023.

Spending subject to appropriation: CBO estimates that five bills would increase spending subject to appropriation (see Table 2). Any spending would be subject to the availability of appropriated funds.

Table 2. Estimated increases in spending subject to appropriation under health care legislation, as ordered reported by the house committee on energy and commerce on december 6, 2023

	By fiscal year, millions of dollars—						
	2024	2025	2026	2027	2028	2029	2024–2029
H.R. 1797:							
Estimated Authorization	*	1	1	1	1	2	6
Estimated Outlays	*	1	1	1	1	2	6
H.R. 2365:							
Estimated Authorization	*	1	*	1	*	1	3
Estimated Outlays	*	1	*	1	*	1	3
H.R. 4310:							
Estimated Authorization	*	*	*	1	*	1	2
Estimated Outlays	*	*	*	1	*	1	2
H.R. 5202:							
Authorization	5	5	5	5	5	0	25
Estimated Outlays	4	4	4	5	5	0	22
H.R. 6132:							
Estimated Authorization	*	1	*	1	*	1	3
Estimated Outlays	*	1	*	1	*	1	3

* = between zero and $500,000.

H.R. 1797, the Setting Consumer Standards for Lithium-Ion Batteries Act, would require the Consumer Product Safety Commission (CPSC) to issue a final safety standard to reduce the risk of fire from rechargeable lithium-ion batteries that are used to power electric-assist bicycles and electric scooters, for example. Based on information provided by the commission, CBO expects that CPSC would need less than two employees for the first two years after enactment and six employees thereafter, at an average annual cost of $190,000 per employee, to issue and enforce the standard. In total, CBO estimates that it would cost $6 million over the 2024– 2029 period for CPSC to implement H.R. 1797, assuming appr priation of the necessary amounts.

H.R. 2365, the Dr. Emmanuel Bilirakis National Plan to End Parkinson's Act, would require HHS to establish an advisory council and to create and update several plans and reports as part of a national project to prevent, diagnose, treat, and cure Parkinson's disease. Using information about similar activities, CBO expects that HHS would need two employees for the first year after enactment and three employees thereafter, at an average annual cost in 2024 of $160,000 per employee, to carry out activities required under the act. In total, CBO estimates that it would cost $3 million over the 2024–2029 period for HHS to implement H.R. 2365, assuming appropriation of the necessary amounts.

H.R. 4310, the Youth Poisoning Protection Act, would ban the sale of consumer products containing 10 percent or more of sodium nitrite by weight. Using information from CPSC, CBO expects the commission would need less than one employee for the first two years after enactment and around two employees thereafter, at an average annual cost of $190,000 per employee, to enforce the standard. In total, CBO estimates it would cost about $2 million over the 2024–2029 period for CPSC to implement H.R. 4310, assuming appropriation of the necessary amounts.

H.R. 5202, the Virginia Graeme Baker Pool and Spa Safety Re-authorization Act, would authorize the appropriation of $5 million annually over the 2024–2028 period for CPSC to continue a grant program and public outreach concerning the safety of children in pools and spas. The bill would require CPSC to extend grant eligibility to nonprofit organizations, appoint a Director of Drowning Prevention, and report to the Congress annually on the program's results. Using information from CPSC, CBO estimates that the cost of implementing the bill would be $22 million over the 2024–2029 period, assuming appropriation of the necessary amounts.

H.R. 6132, the Awning Safety Act of 2023, would require CPSC to issue a final safety standard for retractable awnings. Using information from that agency, CBO expects the commission would need an average of two employees per year, at an average annual cost of $190,000 per employee, to issue and enforce the standard. In total, CBO estimates it would cost about $3 million over the 2024–2029 period for CPSC to implement H.R. 6132, assuming appropriation of the necessary amounts.

Pay-As-You-Go considerations: The Statutory Pay-As-You-Go Act of 2010 establishes budget-reporting and enforcement procedures for legislation affecting direct spending or revenues. The net changes in outlays for the 10 bills that are subject to those pay- as-you-go procedures are shown in Table 1.

Increase in long-term net direct spending and deficits: CBO estimates that enacting H.R. 4881 would increase long-term net direct spending and that such spending would increase by more than $5 billion in at least one of the four consecutive 10-year periods beginning in 2035.

CBO estimates that none of the other bills discussed in this estimate would increase net direct spending or deficits in any of the four consecutive 10-year periods beginning in 2035.

Mandates: H.R. 1797 would impose a private-sector mandate as defined in the Unfunded Mandates Reform Act (UMRA) by requiring manufacturers of electric-assist bicycles and electric scooters, for example, to comply with a prospective CPSC safety standard concerning the risk of fire in lithium-ion batteries. Limited data are available about the extent of industry compliance with the current voluntary standards or about the cost of bringing products into compliance. Therefore, CBO cannot determine whether the cost of the mandate would exceed the private-sector threshold established in UMRA ($200 million in 2024, adjusted annually for inflation).

H.R. 1797 would not impose any intergovernmental mandates.

H.R. 4310 would impose a private-sector mandate as defined in UMRA by banning the sale of consumer products containing 10 percent or more of sodium nitrite by weight. The prohibition would not apply to industrial uses or to food preservation. Because there is only a small market for consumer products containing more than 10 percent by weight and some states already have curtailed the sale of products containing sodium nitrite, CBO estimates that the cost of the mandate would not exceed the private-sector threshold established in UMRA.

H.R. 4310 would not impose any intergovernmental mandates.

H.R. 6132 would impose a private-sector mandate as defined in UMRA by requiring awning manufacturers to comply with a prospective CPSC safety standard concerning fixed and freestanding retractable awnings. CBO expects that the standard could require awnings to be equipped with safety clips and to issue visual or audible alerts when in motion. Based on the cost of such additional equipment and the number of such awnings likely to be sold, CBO estimates that the cost of the mandate would not exceed the private-sector threshold established in UMRA.

H.R. 6132 would not impose any intergovernmental mandates.

CBO has determined that none of the other bills in this estimate would impose intergovernmental or private-sector mandates as defined in UMRA.

Estimate prepared by Federal costs: Austin Barselau (Medicare), Ezra Cohn (public health), Cornelia Hall (Medicare), Hudson Osgood (Medicare), Lara Robillard (Medicare), Sarah Sajewski (Medicare), Katie Zhang (public health), Noah Zwiefel (Medicare); Mandates: Andrew Laughlin.

Estimate reviewed by: Sean Dunbar, Chief, Low-Income Health Programs and Prescription Drugs Cost Estimates Unit; Kathleen FitzGerald, Chief, Public and Private Mandates Unit; Sarah Masi, Senior Adviser, Budget Analysis Division; Asha Saavoss, Chief, Medicare and Health Systems Cost Estimates Unit; Chad Chirico, Director of Budget Analysis.

Estimate approved by: Phillip L. Swagel, Director, Congressional Budget Office.

Federal Mandates Statement

The Committee adopts as its own the estimate of Federal mandates prepared by the Director of the Congressional Budget Office pursuant to section 423 of the Unfunded Mandates Reform Act.

Statement of General Performance Goals and Objectives

Pursuant to clause 3(c)(4) of rule XIII, the general performance goal or objective of this legislation is to direct the CPSC to promulgate a final consumer product safety standard to protect consumer lives and property against the risk of fires caused by lithium-ion batteries.

Duplication of Federal Programs

Pursuant to clause 3(c)(5) of rule XIII, no provision of H.R. 1797 is known to be duplicative of another Federal program, including any program that was included in a report to Congress pursuant to section 21 of Public Law 111–139 or the most recent Catalog of Federal Domestic Assistance.

Related Committee and Subcommittee Hearings

Pursuant to clause 3(c)(6) of rule XIII, the following related hearing was used to develop or consider H.R. 1797:

- On September 27, 2023, the Subcommittee on Innovation, Data, and Commerce held a hearing on H.R. 1797. The title of the hearing was "Proposals to Enhance Product Safety and Transparency for Americans." The Subcommittee received testimony from:
  - Kathleen Callahan, Owner, Xpertech Auto Repair;
  - Scott Benavidez, Chairman, Automotive Service Association;
  - Steven Michael Gentine, Counsel, Arnold & Porter, LLP;
  - John Breyault, Vice President of Public Policy, Telecommunications and Fraud, National Consumers League; and
  - David Touhey, Principal, Connett Consulting, appearing on behalf of International Association of Venue Managers.

Committee Cost Estimate

Pursuant to clause 3(d)(1) of rule XIII, the Committee adopts as its own the cost estimate prepared by the Director of the Congressional Budget Office pursuant to section 402 of the Congressional Budget Act of 1974.

Earmark, Limited Tax Benefits, and Limited Tariff Benefits

Pursuant to clause 9(e), 9(f), and 9(g) of rule XXI, the Committee finds that H.R. 1797 contains no earmarks, limited tax benefits, or limited tariff benefits.

Advisory Committee Statement

No advisory committees within the meaning of section 5(b) of the Federal Advisory Committee Act were created by this legislation.

Applicability to Legislative Branch

The Committee finds that the legislation does not relate to the terms and conditions of employment or access to public services or accommodations within the meaning of section 102(b)(3) of the Congressional Accountability Act.

Section-by-Section Analysis of the Legislation

Section 1. Short title

Section 1 provides that the Act may be cited as the "Setting Consumer Standards for Lithium-Ion Batteries Act."

Section 2. Consumer product safety standard for certain batteries

Section 2 requires the Consumer Product Safety Commission to promulgate a rulemaking under 5 U.S.C. 553 for a final consumer product safety standard for rechargeable lithium-ion batteries used in micromobility devices, including electric bikes and electric scooters, and any related equipment used with such batteries within the jurisdiction of the Commission. Such a standard will be treated as a consumer product safety rule promulgated under section 9 of the Consumer Product Safety Act (15 U.S.C. 2058).

Changes in Existing Law Made by the Bill, as Reported

This legislation does not amend any existing Federal statute.

Index

N

P

R

S

T

V

W